Community Air Screening How-To Manual

A Step-by-Step Guide to Using a Risk-Based Approach to Identify Priorities for Improving Outdoor Air Quality

 United States
Environmental Protection Agency

EPA 744-B-04-001

Disclaimer

Please note that any mention of a trade name or commercial product in this Manual does not constitute an endorsement by the U.S. Environmental Protection Agency

Document title and information for requests or citing: Community Air Screening How-To Manual, A Step-by-Step Guide to Using Risk-Based Screening to Identify Priorities for Improving Outdoor Air Quality, 2004, United States Environmental Protection Agency (EPA 744-B-04-001), Washington, DC.

INTRODUCTION

The Community Air Screening How-To Manual is a resource developed to assist communities in their efforts to understand and improve local outdoor air quality. The Manual is one of a number of resources and tools that are being made available as part of the Agency's Community Action for a Renewed Environment (CARE) program. Launched in the Fall of 2004, CARE is a program designed by the U.S. Environmental Protection Agency (EPA) to help communities work at the local level to address the risks from multiple sources of toxics in their environment. CARE promotes local consensus-based solutions that address risk comprehensively.

The Community Air Screening How-To Manual includes information on how to develop a detailed source inventory, estimate ambient air concentrations, and use a risk-based screening process to identify priorities. The Manual also includes information on consensus building in a community which can be useful in mobilizing local resources for voluntary actions to reduce risk.

The How-To Manual is based on several community-based pilot programs and is the product of broad Agency collaboration. A formal external peer review of the How-To Manual was conducted by eighteen expert reviewers representing a broad spectrum of stakeholders. The draft final of the How-To Manual is now available for public review on the EPA website at:

http://www.epa.gov/oppt/cahp/howto.html

For more information on CARE and the many other tools and resources available to communities, visit www.epa.gov/CARE

Table of Contents

Appendices

List of Figures

List of Tables

How to Use This Manual

Why was this Manual written?

This Manual is designed to support the new efforts that communities and neighborhoods across the nation are making to improve the environmental quality and health of the places where they live and work. Efforts to understand and improve air quality are a key part of this work to create healthy local environments. But many communities have found the work to understand and improve local air quality to be complex and difficult. Even regulatory agencies with expertise may have difficulty answering community questions regarding the combined impacts of releases from multiple sources. While some regulatory programs may consider combined impacts, permitting programs have usually been set up to deal with facilities on an individual basis and they cannot be easily used to answer community concerns about releases from multiple sources. Similarly, the available data on air quality, designed to support national, state, and tribal regulatory programs, is often too general to address the details of local air environments. This Manual was written to help communities overcome these difficulties and find answers to some of the basic questions about air quality that they have not been able to answer in the past.

What does this Manual contain and how is it organized?

There are many different approaches that communities can take to get a better understanding of local air quality. Finding the approach that best matches your community's resources and goals will be a key to the success of your effort to improve air quality. Most of this Manual provides a detailed guide to using **risk-based screening**, one of the

Organize
Collect
Analyze
Mobilize

approaches to analyzing air quality. Chapters 3 through 12 of this Manual contain a step-by-step guide for using the risk-based screening approach.

If you are interested in learning more about the risk-based screening method, please go immediately to Chapter 3 for an overview of this method. If you decide that risk-based screening will help you reach your goals, then this Manual will provide a lot of the assistance that you will need.

If you don't know which approach would be best for your community to use to get a better understanding of local air quality or if you are just getting started in an effort to improve air quality, then Chapter 2 is designed to help you. Choosing the best approach for your community to take to understand your air quality and doing effective work to improve community air quality require **a partnership with adequate resources and a clear idea of what it wants to accomplish**. Chapter 2 explains how to form the partnership that you will need, clarify your goals, and, then, choose the approach for analyzing local air quality that best matches your resources and goals. If, after you have considered the different approaches outlined in Chapter 2, you decide that risk-based screening does meet your needs, then the remaining chapters of this Manual will be helpful. If you decide to use a different approach, then the Manual will refer you to other sources of information for help. Even if you use a different approach, you may still want to use some parts of this Manual as a resource for your work.

Find the approach that best matches your community's resources and goals

As an individual with concerns about air quality, how could I use this Manual to start working to improve air quality in my community?

If you are an individual or a member of a community group or organization and you are interested in doing something to improve local air quality, you can use Chapter 2 to help you get a partnership started. Or, if you are not sure if there is enough interest in air quality to form a partnership, you, or you and your organization, could first follow one of the simpler approaches for collecting available information on local air quality described in Chapter 2 and use this information to help you better understand and explain the need to improve air quality. This will help you to get others interested in joining a partnership. Once you have a partnership and you have worked together to clarify your goals, you can study the options for understanding air quality presented in Chapter 2 (and others that you can learn about) to decide which approach to take to improve local air quality.

We already have a partnership formed to work on air quality or another community health concern, so how could this Manual help us?

If you already have a partnership formed that is interested in learning about and improving local air quality and you are clear on your goals, then you can go directly to the section in Chapter 2 that describes different options for understanding air quality to get help finding an approach that matches your goals and needs. We recommend reading the sections on partnership building and clarifying goals even if you already have a partnership and goals, since these sections raise issues and lessons learned from other partnership efforts to improve air quality that may help you strengthen your work.

If, after considering the alternatives, your partnership decides that the risk-based screening approach meets your needs, you can turn to Chapter 3 and use the help provided there and in the subsequent chapters to get started and complete a risk-based screening project.

Partnerships make things possible

If this Manual and all the work and resources needed to improve air quality look a bit overwhelming, how can I or my organization get started?

As you begin to go through this Manual, one of your first reactions will probably be that the effort required to improve air quality is very complicated and looks as if it would require a whole lot of resources that you do not have. As a resident, the science and the scientific terminology may appear beyond your reach. If you are a professional working in government, industry, or academics, the work in a community may appear equally overwhelming. But before you abandon the idea of working on your air quality, consider the following.

Partnerships make things possible. With a partnership, tasks that seem impossible can get done. The truth is that none of us as an individual working alone has a chance to improve air quality. It is a complicated task that requires the efforts of a lot of people. But looking at the work to improve air quality as either an individual task you must do or as a task that someone else should do may be the problem. Like many of the things that are most important to us, improving air quality is a job that can only be accomplished by everyone working together. All the resources you need to improve air quality are available to your community. There are willing people with all the skills you need to complete the job. Even if the level of trust among residents, governments, and industry is not good, there are steps you can take and resources you can use to begin to change this. So don't let the size or any part of the task intimidate you. Once you get started you will find the help and resources that you need.

You may also worry about starting an effort that will end up being a draining experience for you, with countless meetings and a work load that you are not sure that you could handle. But partnership projects do not have to be draining. Of course they take work, but you will be

Meeting of the community partners of the St. Louis North Side Clean Air Project.

able to divide up the work to make it reasonable. And it will be up to you to set the schedule so you can, and should, set a reasonable pace for you and your community. Working to improve air quality can be a great opportunity to build your community, to meet new people, to learn new things, and to develop new relationships that your community will need in the future. Communities tackling important issues and building partnerships to find solutions are the essence of our democracy at work. Instead of seeing the work to improve air quality as draining, look at it as important work that will be challenging, exciting, and doable.

And you will not be alone. Community organizations and partnerships across the nation have begun to tackle the problem of local air quality, so you will have tremendous resources to draw on to help you make progress. What looks like an impossible task for a single community can look manageable if it is viewed as a joint effort of many communities working together. So plan time to learn about and share ideas with other communities working on similar air quality concerns. Watch for and plan to attend the national or regional meetings that are organized to bring people together to share ideas.

Finally, getting something done in a community always requires a few champions or leaders who can get things started and keep them going. Champions for these projects can come from anywhere. The only qualification seems to be a willingness to learn and a determination to make things happen. A resident, a leader of a community organization or church, an engineer in a local company, a small business owner, a local teacher, a government staffer—anyone can be a champion for an effort to improve local air quality. If you or your organization are willing to champion a partnership effort, you will need patience and persistence, since it takes time to accomplish something new, but, if you persist and don't get discouraged when you run into the inevitable problems that come with any new effort, you will find that your persistence will pay off. You will find that people do want to work together and that we can explain things to each other and make decisions together. And you will find that the process can be exciting, challenging, and rewarding. Most importantly, you will succeed in improving your air quality and you will make a real contribution to the health of your community and the health of your community's children. And when you have made some progress, share your experiences. In fact, you and your partners may find a way to turn your experiences into the next and improved version of this Manual.

Communities tackling important issues and building partnerships to find solutions are the essence of our democracy at work

Getting Started:

Building a Partnership
Setting Goals
Choosing a Plan to Meet
   Community Goals
Developing a Communication Plan

Organize
Collect
Analyze
Mobilize

How is this chapter organized and what does it cover?

This chapter is divided into four parts to discuss important aspects of the work needed to get started on a project to understand and improve local outdoor air quality. The first part on building a Partnership describes the work needed to get all the people and resources together that you will need to improve local air quality. It also describes ways of organizing the Partnership that will help to build the trust you will need to work together effectively. The next section of the chapter describes the work you will need to do to clarify the goals of your Partnership. This section describes the need for all the partners to explain what they would like to accomplish and then work together to find the goals that everyone can share. The third section on choosing a plan for work describes some of the different approaches that you could take to reach your goals. Once you have formed your Partnership and clarified your goals, you will be in a position to discuss these different approaches and find the one that best matches your goals and the resources that you have available. Finally, the chapter includes a section on developing a communication plan. This section talks about the important work needed to communicate with and involve the broadest possible number of community residents and businesses in the work of the Partnership. It also discusses the importance of communicating with all the key stakeholders outside the community to ensure that they understand and support the work you are doing.

What kind of assistance does this Manual provide to help with building a partnership, clarifying goals, and developing a communication plan?

This Manual focuses primarily on providing the help a community will need to use risk-based screening to understand and improve local air quality. The Manual provides only general information and a list of resources for the work of building a Partnership, clarifying goals, and communicating with the community. The Manual focuses on the technical aspects of risk-based screening because help for this part of an effort to improve air quality is not currently available to communities—not because the non-technical parts are less important or require less effort. This chapter of the Manual discusses only those issues in the areas of building the Partnership, clarifying goals, and communicating with the community that are particular to efforts to improve local air quality. For a more complete discussion of these topics, communities should supplement this Manual with other resources that are available to help communities in these areas. A list of resources for building a Partnership, clarifying goals, and communicating with the community and other stakeholders can be found in Appendix A.

Building a Partnership

What is the value of a Partnership?

Building a broad, working Partnership is key to the success of the work. The effort needed to get a better understanding of and improve local air quality is complex and will require a wide range of skills and resources. No single sector of the community or level of government has the ability or resources to do this work alone. Only a Partnership will have the ability to bring together the resources, information, and skills that will be needed. To improve air quality, a community must reach an agreement on an effective plan for action and then work together on its implementation. The Partnership and the work to build the Partnership will also provide the means for different parts of the community to share ideas and develop the trust that will be necessary for joint action.

How will a Partnership for improving local air quality get started and what would it look like?

The strategy for getting a Partnership started will be different for each community. The approach will depend on factors such as the kinds of organizations that already exist in a community, the ability to access technical resources, and the local interest in air quality issues. Whatever the situation, forming a Partnership will require an organization or individuals to take the lead and act as a consistent champion for the idea of working together to improve air quality. In most cases, the potential members of a Partnership will have little experience working together to address air quality issues, so it will take time and consistent leadership to get started.

The leaders who champion this effort will start to form a Partnership by convincing others of the value and the potential of working together to improve local air quality. The Partnership may be formed as a part of, or separate from, existing community organizations. If it is possible, using existing organizations with their infrastructure and established ties will save the Partnership from the need to build an organization and develop ties to the community from scratch. A Partnership to improve air quality could be viewed as an inclusive community organization with several levels of involvement. A core Partnership group of somewhere around 20 members would have the responsibility to lead, organize, and carry out the work needed to understand and improve air quality. A much larger group of community members would participate occasionally in the

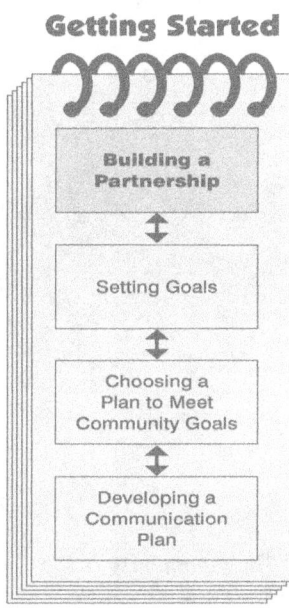

Getting Started

Building a broad,

working

Partnership is the

key to the success

of the work

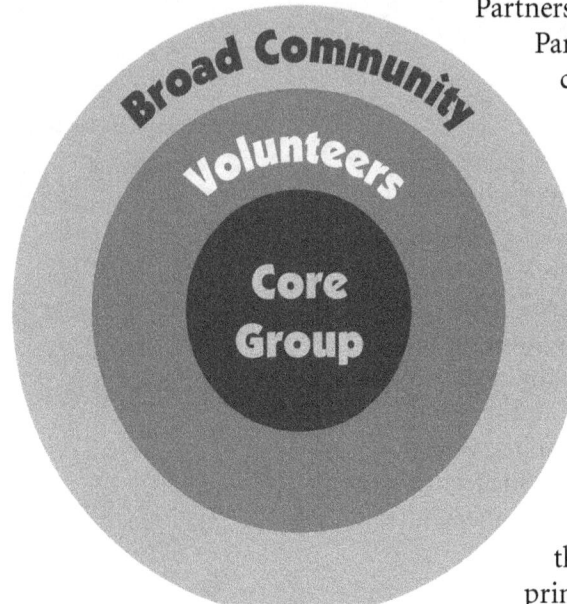

The community is the Partnership

Only a Partnership that can include and mobilize large parts of the community is likely to have the resources to improve air quality

Partnership by joining in activities organized by the core Partnership group. These activities could include collecting information on sources and helping to communicate results to the community. At the broadest level, the entire community can also participate in the Partnership by attending public meetings, providing input, and taking part in community mobilizations to improve air quality. Only a Partnership that can include and mobilize large parts of the community is likely to have the resources to improve air quality.

Because the work to address air quality requires broad participation of all sectors of the community, it may help for the members of the core Partnership group to view themselves primarily as a community leadership group and not as an independent organization trying to tackle air quality issues by itself. As community leaders, the job of the core Partnership will be to help all sectors of the community develop a better understanding of air quality, set clear and realistic goals, and mobilize to take the actions needed to improve air quality. Adopting the perspective of the core Partnership as leaders of the whole community will encourage the core Partnership group to organize its work so that it includes as much contact and interaction with the broader community as possible.

In forming the core group of the Partnership, it will be important to include a balanced representation from as many different sectors of the community as possible. Community members who have been active around air quality issues, if there are any, will be key members, but the core group should also include representatives from groups in the community not currently involved in air quality issues. This broad representation from all parts of the community will be key to ensuring that all views will be considered and that the Partnership will have access to the information and the support that it will need to reach its goals. It will also be important to include members in the core group that have the skills and resources that will be needed to complete the project. A list of those skills is included in this section.

Who might participate in the work of a Partnership to improve local air quality?

The Partnership will be made up of a combination of members from all sectors of the community and representatives of governments and organizations interested in supporting the community effort to improve air quality. If possible, the majority of the Partnership core group should be made up of people who live or work in the community. Community residents, community health professionals, community businesses, and other community members have the most stake in the immediate and long-term health of the community. They also will be the key source for the energy and resources that will be needed to improve air quality.

Members of the Cleveland Clean Air Century Campaign working group and EPA at the launch of the partnership.

To a certain extent, the membership of the Partnership will depend on the goals it establishes. The development of the Partnership and the clarification of its goals (discussed in the next section) are inseparable. The addition of new partners may change goals and, similarly, the refinement of goals may influence the Partnership's composition. For example, if a community identifies work on a particular health concern related to air quality, such as asthma, as a goal, the Partnership might expand to include members, individuals, and organizations with the resources and skills needed to address this issue, such as the American Lung Association, local health professionals, teachers or staff from local schools, and so forth. Depending on their goals, Partnerships may use the following "checklist" of the kinds of organizations that might be considered as recruitment pools for membership in a local Partnership to improve air quality.

Places to Look for Partners

✔ Community residents

✔ Community civic, environmental, and economic development organizations and associations

✔ Housing associations

✔ Churches

✔ Teachers and staff at public and private community schools

✔ Community students and student organizations, including environmental clubs

✔ Youth organizations such as 4-H and Scouting

✔ Local library staff

✔ Local businesses and industry

✔ Local business associations

✔ National business associations

✔ Unions representing local employees

✔ Colleges and universities

✔ College students and student organizations

✔ Local government, including elected officials and agency representatives from health, environmental, planning, permitting, development, public works, parks, police, and fire departments

✔ State and tribal governments, including transportation, environment, health, and natural resources departments

✔ Federal government agencies, including environment, housing, energy, transportation, forestry, etc.

✔ National, state, and tribal environmental organizations

✔ Environmental justice organizations

✔ Public health organizations

✔ Local foundations concerned with the environment, public health, or community development

What effort will it take to build a Partnership?

The amount of work required to build a Partnership around air quality will depend, in part, on the existing relationships among the potential partners. If a strong community organization with good ties outside the community takes the lead, the work to form a Partnership may not take long. If, however, the level of trust among potential partners is low, and there has been a history of contention over air quality issues, the work to form a Partnership will require significant time and effort. In Baltimore, for example, where the level of trust was low, the effort to form a Partnership took a year. Whatever the situation in a community, the work to develop a Partnership is as important to the success of the project as the technical work to identify priorities and find solutions. Efforts invested at the outset and throughout the work to build and maintain a successful Partnership will pay off in results obtained later in the process. It will be especially important to remember that partnership building is an effort that will need to be maintained for the length of the project. Partnership building is not just a task for the beginning of an effort. To sustain itself, the Partnership and its members will need to do all their work in a way that continues to build the Partnership throughout the course of the work to improve local air quality.

What needs to be done to ensure that all members of the Partnership participate as full and equal partners in the process?

Make sure that all members have the opportunity to be heard and to participate fully as equals in the work and decisions of the Partnership

The Partnership will be most effective if it makes sure that all of its members have the opportunity to be heard and to participate fully as equals in the work and decisions of the Partnership. Input from individuals and from leaders of community and business organizations, schools, and churches will help to ensure that all viewpoints are considered and that the Partnership has access to a wide range of community resources. But building a true partnership that can realize the full potential of the community may require an effort to overcome some obstacles. Organizations that are used to making decisions will need to learn to share decision-making with residents, small community businesses, and community organizations. And methods of conducting Partnership business that can discourage participation of community members, such as using e-mail to communicate between meetings or using professional terms and technical jargon during meetings, will need to be avoided. And even though all sectors of the community have equally important contributions to make to the effort, the opportunity to participate may not be equal. Volunteer residents will normally participate in Partnership work in the evenings after work. These volunteers may not have as much time to devote to Partnership work as government and industry participants, so holding too many meetings may discourage participation. The participation of some volunteers may also be limited because they do not have an adequate background in the science used to understand air toxics and to identify priorities. Similarly, government and business Partnership members who do not live or work in the neighborhood will be limited by their lack of knowledge or understanding of the local area. All these and other similar obstacles to creating an effective and real partnership should be acknowledged and considered at the beginning of the Partnership.

To compensate for these differences in resources and backgrounds, the Partnership's organization and work can be designed in a way that will promote the fullest possible participation of all of its members. This can be accomplished by means such as by arranging meetings to accommodate participants' available time and schedules, by taking the time to share information and provide any necessary background, by operating by consensus, and by sharing or rotating the leadership of the Partnership. In addition, funding for a person to provide staff support for committee members who are volunteers and to facilitate overall community participation may be necessary to ensure the full participation of all community residents.

Making decisions by consensus in a partnership can help to build trust

Making decisions by consensus in the Partnership can help to build trust and ensure that all the partners participate fully. In an organization based on consensus, decisions require the unanimous support of all participants. Since decisions require everyone's support, partnerships using this approach must ensure that everyone's views are fully considered and accounted for. Taking a consensus approach will encourage the Partnership to discuss issues thoroughly and to search for an approach that meets the common good of the Partnership and the community as a whole. Decisions reached by consensus may not perfectly match all the needs of each partner, but they will generally represent an improvement that everyone can live with and that moves the community as a whole closer to the goals it holds in common.

It is also important for the Partnership to have technical members with diverse backgrounds to ensure that the scientific advice and training the Partnership relies on to make decisions is sound and unbiased. This can be accomplished by bringing technical expertise to the Partnership with a variety of backgrounds, such as technical residents of the community, community health providers, local science teachers, technical staff from governments and industry, and faculty and students of academic institutions. If the Partnership cannot find sufficiently diverse sources for technical advice, funding to support an independent scientific adviser may be required.

What will the core members of the Partnership be expected to do?

At a minimum, to develop an understanding of local air quality and to carry out a plan for improvements, each of the core members of the Partnership will need to be willing to do the tasks in the following checklist.

Partnership Responsibilities Checklist

✔ Fairly and honestly represent the views of the community residents, businesses, and organizations in Partnership discussions and decisions

✔ Share information so that all Partnership members have the understanding necessary to participate fully in the work

✔ Listen carefully and consider fairly the views of other members of the Partnership and work to develop a collaborative decision-making process and to build consensus

✔ Participate in the direction and work of any technical analysis needed by the Partnership and make sure that the technical work considers all appropriate information and is done in a way that is technically sound

✔ Consider the new information developed by the Partnership and use the information as a basis for Partnership decisions

✔ Help to regularly communicate the work of the Partnership to all sectors of the community to solicit their input on the direction of the work and to keep them informed as work progresses

✔ Help to develop and lead the implementation of an action plan to make improvements in air quality

✔ Help with group logistics such as organizing, chairing, and keeping records of Partnership meetings

Because the scope of Partnership activities will depend on the goals that are chosen, the tasks that will need to be completed by the core Partnership members will be developed as those goals are clarified. Once the goals are set and the scope of the effort defined, it will be important to clarify the roles and expectations of core Partnership members so that they can plan to meet their commitments. For the staff of governments and large businesses participating in the Partnership, formal written agreements on roles and responsibilities may be helpful. These written agreements can provide staff with a means of bringing the work of the Partnership to the attention of their organizations.

What skills will the members of the core Partnership group need?

In forming a partnership around air quality, it will be important to consider the skills and tasks that will be needed over the course of the work. As many people as possible who can provide these skills should be included in the core Partnership group. In cases where members with the necessary skills cannot be found, a partnership may need to find funding to provide for these skills or to provide training to Partnership members to develop the skills.

To get the Partnership started on the right foot and to help ensure that all partners have the skills and understanding needed to work effectively in a partnership, it may be a good idea to set aside time at the start to organize training for the core Partnership members in consensus- and team-building skills. Time will also need to be set aside to discuss and decide on key process questions such as ground rules and organization for the Partnership. Providing this training and organizing these discussions will help to ensure that the day-to-day work of the Partnership is done in a way that builds trust and strengthens the Partnership.

The following is a list of skills that will be needed to complete the work outlined in this Manual.

Leadership: Leadership is probably the single most important skill needed to mobilize a community to work together to improve air quality. Successful completion of all aspects of the work will depend on leaders with a clear understanding of the Partnership's goals and direction. Because the Partnership will include a broad range of participants, leadership will be needed from each of the different sectors of the community represented in the Partnership. A core group of community, business, academic, tribal, and government leaders committed to making the Partnership succeed is the key to success.

Willingness and ability to exchange information and to learn from others: These skills will be important for all members of the core Partnership group. Skills in communicating the science used in the screening analysis to nonscientists will be especially important to the Partnership.

Ability to collect information and data on local sources: Government staff, members of environmental organizations, academics, and others familiar with available data, plus community residents and representatives of local businesses familiar with local sources and their releases, will be essential to the collection of information for the screening analysis.

Technical and scientific skills needed for analysis: Skills needed, depending on the goals and work plan set by the community, may include risk assessment, air dispersion modeling, exposure assessment, database management, toxicology, health care, transportation planning, environmental engineering, and pollution prevention. Potential sources for these skills include government and industry staff, college and university faculty and students, local science teachers, local health professionals, residents working in professional fields, and the technical staff of environmental organizations.

Communication skills needed for soliciting input and reporting the work and results to the community: Because the work of the Partnership depends on community input, support, and participation, the ability to explain the work of the Partnership to the community and to solicit community input will be essential to the success of the Partnership. This will require both communication skills and knowledge of the community. Members of community organizations, risk communicators, teachers, community leaders, librarians, and journalists can work together to develop ways to collect community views and to produce reports, press releases, newsletters, and other means of communicating with the community.

Organizational skills: Chairing meetings, keeping records, organizing community events and actions, developing budgets and handling and raising funds, and other related administrative skills will be needed over the course of the project. Some Partnership members should have these skills, and training can be organized for other members to meet the Partnership's needs.

Facilitation skills: The ability to foster a process that will build trust, improve communication, clarify goals, and develop participation in the Partnership will be essential to success. These facilitation skills may be found, or developed, in Partnership members, or professional facilitation can be retained to support the Partnership's work.

Ability to develop and implement plans for making improvements: Finding solutions to air quality concerns will require the active participation of the business community, transportation planners, and community leaders. Environmental engineers and pollution prevention specialists from government, industry, and environmental organizations will also be needed to help identify solutions. Community and business leaders will be needed to help mobilize residents and businesses to implement the plans for making improvements.

The Partnership will provide opportunities for a wide range of activities for community volunteers

What level of commitment will be expected of members of the Partnership?

The core group needed to direct and implement the work of the Partnership will need to meet regularly for the entire course of the project. The frequency of meetings, and the pace of the project, can be set by the members; for example, meetings of the core group could be held once or twice a month. In addition to participating in regular meetings, members of the core group will be expected to carry out the work of the Partnership between meetings. For example, if the Partnership decides to develop a local inventory of sources, some of the core group members will work with others to gather all the information for the inventory and then present their work for review by the full core Partnership group. The amount of work, per month, for the members of the core group, will depend on many factors, including the time allowed for the project, the availability of staff or volunteer support, and the extent of the work needed to carry out the plan for improving local air quality. Depending on the goals, resources, and pace of the effort, the effort to meet initial community goals could take up to two years to complete.

In addition to the work of the core group, the Partnership will provide opportunities for a wide range of activities for community volunteers. Individuals, schools, businesses, and community organizations will be needed to survey traffic, identify sources of chemical releases, develop outreach materials, distribute information, and help to organize and participate in community meetings and activities.

What kinds of things might the Partnership need to support its work?

Depending on the situation in a community, funding to hire a person to work for the Partnership may be helpful or, in some cases, necessary. An active partnership will require a good deal of work to help organize and support all the volunteer activities that will be needed to carry out its work. Staff from existing community organizations, governments, and businesses, or community volunteers may be able to provide this support, but, if this is not possible, funding to hire a full- or part-time staff person to support the Partnership may be necessary to facilitate this work. It may be especially important for the Partnership staff person to focus on developing and maintaining the involvement of the community. The staff person could support resident volunteers on the core committee and work to keep the community at large informed and involved in the Partnership's work.

Funding for professional facilitation may also be helpful or necessary. Working in a Partnership, especially in communities starting with a low level of trust among partners, can be very difficult. The ability to have a trained facilitator focused on partnership building, and on the process of the Partnership, can be very helpful and, in some cases, necessary. Core committee members may have neither the facilitation skills nor the time to focus on the facilitation that will be needed to ensure the success of the Partnership. It is also difficult for Partnership members to represent the views of their community or organization while also serving as facilitator for the Partnership.

As mentioned above, depending on the availability of technical resources, funding for an independent technical advisor may also be necessary to assure all Partnership members that they have the balanced scientific advice needed to make sound decisions.

The Partnership will also need to find a way to obtain or develop the following to support its work:

- Access to meeting space for Partnership committee meetings and for larger community meetings

- Access to a relatively new desktop computer capable of handling the database used to store information collected by the Partnership and for the dispersion modeling if the Partnership decides to conduct its own analysis of local air quality

- A location in the community for the computer and for storing the records of Partnership meetings and decisions

- Equipment or access to equipment to copy, print, and distribute information about the Partnership's work and results
- An effective method for members of the core Partnership group to communicate with each other between meetings

Are there things the Partnership can plan in advance that will help to ensure the success of the efforts?

Planning ahead may be difficult since the entire Partnership is likely to be developing its understanding as the work progresses. But some planning may be helpful. One thing you might consider is establishing a team to begin collecting the information you will need to develop a plan for work at the same time as the Partnership is working to clarify its goals. If this information is available when the discussion to clarify goals is completed, the Partnership's work to decide on a work plan could proceed immediately. Please see the discussion of developing a work plan below for a description of the information that this team could collect.

Teachers and students can be a tremendous resource for the Partnership and community

Once your Partnership has developed its plan for work, try to do some advanced planning and preparation so that you are prepared to complete each step of the effort as the work progresses. The Partnership could identify all the teams that it will need to form over the course of the work and organize the teams early so that they can begin to assemble the resources and develop the skills they will need to carry out their tasks. These teams will work for and report to the core Partnership group. As an example, if the Partnership decides to use the risk-based screening approach described in this Manual, the list of teams needed would include:

- Communications Team, described in this chapter
- Emission Source Inventory Team, described in Chapter 4
- Quality Assurance/Quality Control Team, described in Chapter 4
- Concentrations Estimation Team, described in Chapter 5
- Screening-Level Concentration Team, described in Chapter 5
- Recommendations Teams, described in Chapter 8

Partnerships, depending on their goals and plans for work, will have their own needs and resources, so the teams your Partnership needs may differ from these.

In addition, if the Partnership uses the work as an opportunity to build the long-term capacity of the community (see discussion of this issue in the next section), then some advance planning would allow for an early start on providing the training that community members will need to participate in the work of the teams. Advanced planning is especially important for involving local schools and colleges. Teachers and students can be a tremendous resource for the Partnership and community, but getting a school's commitment and providing the training that the teachers will need to incorporate the work into the school's teaching must be planned months in advance. Advance planning is also crucial for fund-raising, so the Partnership will almost certainly need to set up and provide training for a team to work on fund-raising. Organizing a fund-raising team at the start of the work will help to ensure that the Partnership has sufficient funding to achieve its long-term goals.

Setting Goals

Why is it important to clarify goals at the beginning of the Partnership's work?

Members of the core Partnership group will all share the goals of understanding and improving local air quality. But, at the start of a project, participants will almost certainly have different perceptions of those goals. Partnership members are also likely to have other personal objectives not directly related to air quality that they are hoping or assuming will be included in the scope of the Partnership. Adequate time must be spent at the beginning of the process to discuss and understand the expectations of all the participants in order to discover and clarify the goals that can be shared by all. Clarifying goals will also enable the Partnership to develop a plan for work to match its goals and to ensure that the results of the Partnership's work and the expectations of its members are consistent.

The discussion to clarify goals in the core Partnership group should be viewed as part of a broader process of clarifying goals for air quality in the community as a whole. Agreement around shared goals will be essential to uniting the community for the work that will be needed to improve air quality. As a community leadership organization, the core Partnership group can facilitate the discussion of goals in the broader community. Since members of the core group represent different sectors of the community, discussions in the core group can help to clarify community goals, but to ensure

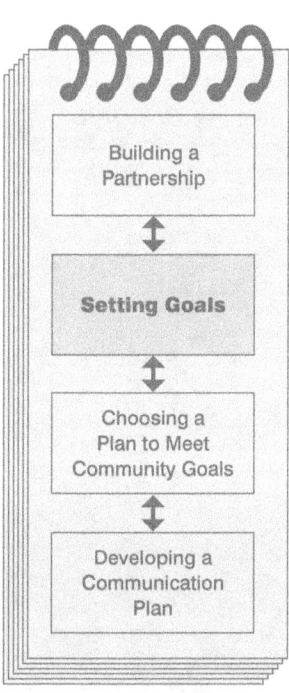

consideration of all views and to ensure that the broader community benefits from the discussion of goals, the core Partnership group should try to organize as much interaction with the community as possible in the process of clarifying goals. For example, core group members could meet to discuss goals with community leaders and organizations, and a large community meeting could be organized to discuss and approve the goals once they are clarified.

As noted earlier, this Manual does not provide detailed guidance on methods communities can use to clarify goals. The discussion of goals in this Manual is limited to issues related to efforts to improve air quality. There are resources available that provide detailed guidance on methods for clarifying goals. Please see Appendix A for a list of the resources available to help organizations establish goals.

It will be very important for core group members to develop an understanding of each other's interests

What are some important issues to consider to set clear Partnership goals?

Identifying members' concerns and interests: It will be important, first of all, to find out why members of the core group and the community sectors and organizations that they represent are interested in air quality. Some Partnership members may be interested in air quality because of their concern for community health or because of a concern about the siting of waste treatment facilities in the Partnership neighborhoods. Other Partnership members may be interested in air quality issues because of the desire to promote economic development, expand community businesses, and revitalize the Partnership neighborhoods. It will be very important for core group members to develop an understanding of each other's interests and to clarify the relationship of the Partnership's goals to the related goals of its members.

Clarifying the relationship of the Partnership's goals to the interests and goals of the participants will also help participants develop realistic expectations for the results of the Partnership's work. For example, air quality is likely to be only one of the factors affecting community health, so the work to improve air quality, by itself, may not be able to meet a member's goal of achieving measurable improvements in overall community health.

Preparing for different outcomes of the work to set priorities: It will be important for the members of the core Partnership group to discuss all the possible outcomes of any analysis done to identify community priorities and what each outcome would mean to each of the members. What if small businesses, large businesses, households, or mobile sources are identified as the priority concern? Or better: What would it mean if my business, my home, or my car was identified as a community priority for potential action?

Many, perhaps most, members of the Partnership will enter the process with a conviction about the sources that will need to be focused on to improve air quality. Different members will have different sources in mind for the community's priority, so the expectations of all the Partnership members for the focus of work to improve air quality cannot be met. A discussion of all the different possible outcomes will allow participants to consider carefully what the project results might mean for them and for their goals.

Setting realistic expectations for accomplishments: To clarify goals, it will also be important to discuss, in detail, what the Partnership will be able to do to improve air quality when the analysis is completed and priorities have been identified. Questions that the Partnership will need to consider include:

- What resources will be available to make changes?

- What issues can be addressed by the local community and which ones, such as requirements for new vehicles or vehicle inspection programs, would potentially require broader action?

- What could be done if the screening analysis identifies a concern from a large business? Or a concern from a small business? Or from mobile sources? Or households?

- In what circumstances would enforcement and regulatory authority be used to improve air quality? And what kind of information will be required to support this approach?

- If the actions will be voluntary, what resources and commitments does the Partnership have to accomplish its goals and, again, what kind of information would best support the voluntary approach?

Discussing possible outcomes and the resources that may be available to address them will help the Partnership set realistic goals for its efforts to improve air quality.

Understanding the relationship of air quality goals to other community priorities: Understanding and improving air quality will not be the only community priority. Most communities will be working on other issues, such as education, jobs, crime, and health. It will be important to identify these other community priorities and the ongoing work to address them, so that the work on air quality can be designed to support and complement these priorities. For example, if a community has developed a plan for improving public health as a part of the Healthy People 2010 national program, it will be important to understand the relationship of the work on air quality to this community effort. With limited time and resources, communities can work on only a limited number of priorities. The ability to integrate work on air quality into the other priorities of a community may be essential to finding the resources that will be needed to address air quality issues.

Understanding other community priorities will also make it possible to organize work to avoid unnecessary conflicts and opposition. For example, if a community is interested in creating more jobs and developing the local economy, including, possibly, the redevelopment of local brownfields sites, it will be important to organize the work of the Partnership in a way that does not undermine those community efforts. By stressing the importance of improving local air quality as an incentive for businesses looking for new locations and organizing the work on air quality in a way that supports the community development goals, the Partnership may be able to achieve its goals without disrupting development efforts.

Identify areas where there is already sufficient agreement in the Partnership to begin immediate work to improve air quality

Consider setting short-term goals to organize immediate action to address known community concerns: Some members of the community and the Partnership will be more interested in action than in studying local air quality. It will be important to identify areas where there is already sufficient agreement in the Partnership to begin immediate work to improve air quality and community and environmental health. Examples of projects that might be started include working with schools, students, and parents to address indoor air problems, helping families address asthma concerns, developing community plans for ozone alert days, or working to provide pollution prevention assistance to local businesses. These concrete efforts to improve air quality and community health will increase awareness and trust in the Partnership's work in the community and set the stage for broader mobilization efforts to improve air quality that the Partnership will organize when the screening analysis is completed. For examples of projects communities have adopted to improve air quality, see Appendix B.

Consider making the building of long-term community capacity to address air quality issues a Partnership goal: A discussion of what it would mean for the Partnership to set a goal to build the long-term capacity of the community to understand and address local air quality issues may help to raise important issues for discussion.

Questions to discuss may include:

- How will it be possible to retain the knowledge and skills learned in the course of the work after the Partnership has completed its work?

- How will the community keep up with future impacts on air quality to ensure that local air quality continues to improve?

Adopting a goal to build the long-term community capacity will require the Partnership to plan its work so that it is done in a way that builds community capacity to address air quality issues. A long-range view of capacity may require more emphasis on developing long-term relationships and trust, more emphasis on using the work to train members of the community, more attention to organizing information so that it can be updated to monitor future changes, and more work to develop a permanent organization in the community that can continue to address air quality issues.

Finding the funding that will be needed to support the future organization and work of the Partnership will be an essential part of building the long-term capacity of the community to address air quality issues. Given the importance of funding, and the amount of work and length of time required to apply for funding, the Partnership may want to make fund-raising a key part of its work from the start. Organizing and providing training to a fund-raising team at the beginning of the Partnership's work would allow enough time for the Partnership to develop the skills and complete the work that will be needed to find sufficient funding to sustain the long-term work of the Partnership.

Plan work so that it is done in a way that builds the community's capacity to address air quality issues

How can the Partnership consolidate its process of clarifying goals?

After a broad consideration of all the issues discussed above and as much interaction with the broader community as possible, the Partnership will be in a position to make some decisions and clarify its goals. This is the time for members and the Partnership to review and finalize goals and commitments. The Partnership will have to decide, in the light of the clarification of the possible outcomes and the relationship of the outcomes to the goals of the members and the community, if its members are still committed to working in a partnership to understand and improve local air quality. The Partnership will also have to decide if the goals of the members and community would be better served by expanding the scope of the Partnership to include other issues in addition to air quality. Because of differing resources and situations, the decision on goals and approaches will be unique to each community.

For example, the Partnership could decide to add resources and expand the work of the Partnership to include a more general approach to community health or economic development. If that is the case, the work on air quality would become a part of a larger community effort addressing multiple issues. The Partnership might also decide that it does have broader goals, but that they are already being addressed by other means in the community and a partnership to address air quality would complement and support the work on the other goals. A community might also decide that, while it does have other goals, it does not have enough resources to address more than one goal at a time. In that case, the Partnership could set long-term goals for the work to improve air quality as one of the steps in a long-term plan.

Communicating the statement of goals will establish the process of soliciting review and input from the community

Completing all this planning and discussion is important because it will make sure that the Partnership's work on air quality will be integrated into the work going on in a community and that all the members of the Partnership have a common and realistic understanding of the work they are about to begin.

Once the Partnership has reached a consensus on its goals, it will be important to consolidate the discussion and decisions by writing a clear statement of its goals. This goal statement should include a summary of the discussions of all the important issues related to the goals. The work to develop a written explanation of Partnership goals will ensure that everyone in the Partnership is clear on the goals of the work. In addition, the written explanation that is produced can be used to communicate with the broader community to solicit their review and input. A written goal statement can also serve as a reminder that the Partnership can use to stay focused as work progresses. The statement of goals could include:

• A statement of the Partnership's goals

• A list of the members of the core Partnership group

• A summary of the work that the Partnership will do to reach its goals

• A discussion of the relationship between the Partnership's goals and the other goals of the members of the core Partnership group

• A discussion of the relationship between the Partnership's goals and the other goals of the community

- A discussion of results that can be expected from the work

- A discussion of any limitations of the work that is to be done

- A discussion of the different possible outcomes of the screening analysis to identify priorities and the realistic expectations for results in improving air quality for each of the possible outcomes

- A discussion of what the Partnership will do to use its work to build the long-term capacity of the community to address air concerns

This statement of goals may be the Partnership's first product available for communication to the community. Communicating the statement of goals will establish the process of soliciting review and input from the community. It will also begin the process by which the core Partnership group summarizes its work to inform and educate the community on local air quality. The statement of goals can be used to develop and implement the communication plan discussed in the final section of this chapter.

Choosing a Plan to Meet Community Goals

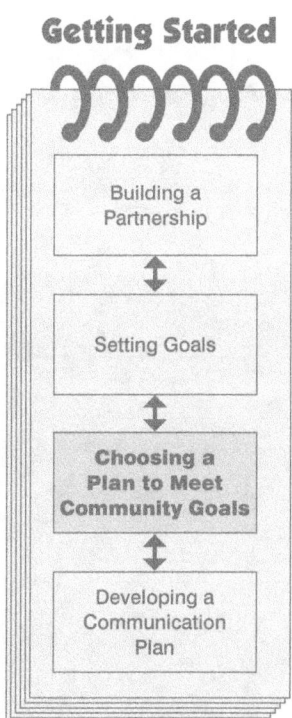

Getting Started

What preparations will we need to develop a work plan to reach our goals?

Once your community has set its goals, you will need to develop a plan for the work you will need to do to reach your goals. Developing an effective work plan may require attention to identifying priorities, finding solutions, and communicating with and mobilizing the community for potential action. Since the Partnership will need to make decisions on all of these aspects of the work plan, this may be an important time for the Partnership to take some time to make sure that all of its members have the background necessary to participate fully in the effort to develop and choose an effective plan for work. This will require Partnership members to exchange information with each other. Residents and business representatives will need to begin sharing their understanding of the community and their insight into how the community, both businesses and residents, can be motivated to take actions to improve air quality. Technical members of the Partnership will need to share their understanding of the science that will be needed to understand air quality and to choose an effective method for identifying priorities for potential community action. Choosing a work plan is crucial to the success of the project, so adequate time should be taken at this point to ensure that all members of the Partnership have all the information they need to participate fully in this choice.

How will this section help us to develop a work plan to reach our goals?

To help you decide on a plan for work, this section of the Manual will describe options for collecting information and carrying out the technical analysis to understand local air quality. This section also provides some examples of approaches that communities have used to carry out this technical analysis. You can use this discussion to help you develop your own plan for work.

Indoor air and acute effects of chemicals in outdoor air are important aspects of community air quality

What aspects of local air quality will be addressed by the work plan options presented in this section?

The options described here focus on understanding the potential **chronic, long-term effects, such as cancer, of chemicals in outdoor air** on community health. If your goals include understanding and improving other aspects of air quality, you will need to consider work plan options beyond those described in this section. Most importantly, options for developing a work plan for understanding or improving **indoor air quality** and for **understanding and addressing the acute, short-term effects of the chemicals in outdoor air** are not included.

Both indoor air and the acute effects of chemicals in outdoor air are important aspects of community air quality. Because people spend much of their time indoors, exposure to chemicals in indoor air can be significant. And many communities have concerns about acute effects due to events such as summer high-ozone days or peak releases from facilities with control equipment failures, or releases due to maintenance or accidents. Understanding these aspects of local air quality can be important to setting community priorities and working to effectively reduce risks due to air toxics. You may want to consider, and we recommend, organizing a team to collect the information and develop the work plan options needed to address these issues. Many of the methods used to address chronic effects described in this Manual could be adapted to address acute effects. For references to sources for information and assistance on indoor air and acute effects of chemicals in outdoor air, please see Appendix C. In the future, we hope to expand this Manual or develop supplemental guides to provide more assistance to communities to address these issues.

In addition to the effects of air quality on the health of the Partnership's community, some communities may also be interested in other important aspects of air quality, such as the impact of local sources on the health of other communities or issues such as visibility or effects on ecological health. These important aspects of air quality are beyond the scope of this Manual, and the Partnership will need to rely on the expertise of its own members or other resources if it wants to address concerns not related to community health.

And one final reminder: Obviously risks from air toxics are not the only health concerns facing communities. Partnerships will want to ensure that all significant environmental and non-environmental health risks, such as childhood lead poisoning, drinking water contamination, and drugs, are being addressed and that work on air quality is done in a way that complements other efforts to improve community health. Some communities with limited resources may have to choose to address some risks before others.

Are there other considerations to keep in mind as we develop our work plan?

As you consider the options presented here and the examples of approaches used in other communities, please remember that the work to understand and improve local air quality is still developing, so no one is really sure of the best way to approach this work. Community partnerships will have to use their best judgment to choose a plan for work and then, if possible, communicate their experiences to other communities so that we can all learn from each other in our efforts to improve local air quality. Other options may be available, so use the approaches presented as a way to start your discussion, not as a way to limit your choices. To help facilitate communication among communities working on local air quality, EPA's Air Office has set up a database with information on completed and ongoing community efforts to improve air quality. This database, the Community Assessment and Risk Reduction Database, is available on the Internet at http://epa.gov/ttn/atw/urban/urbanpg.html.

It may also be helpful to keep a few thoughts in mind as you consider the options for the technical aspects of your work plan. To build consensus and to work effectively to improve local air quality, you will have to avoid two errors: developing a work plan with too much

Partnerships will have to use their best judgment to choose a plan for work

analysis that results in long delays to getting to the work to improve air quality, or too little analysis that results either in not having enough consensus in the community to take action or in taking actions that do not effectively address important community risks. Each community will have to find the balance between analysis and action that results in the most improvement in local air quality. Depending on factors such as the level of agreement in the community and available resources, this balance will differ for each community. And as you develop your work plan, remember also to consider the possibility, discussed in the section on setting goals, of doing the technical parts of the work plan in a way that builds the long-term community capacity to address air quality concerns. If you choose this approach, your work plan will need to include specific plans for building capacity.

Each community will have to find the balance between analysis and action that results in the most improvement in local air quality

And remember, this is the section of the Manual that will help you determine if the risk-based screening method described in Chapters 3 through 12 of the Manual is appropriate for your community. Risk-based screening is one of the options discussed below. If, after considering the options, you choose to use all or part of the risk-based screening approach, then the remaining chapters of this Manual will provide you with detailed assistance.

Also, please note that this Manual focuses primarily on the technical analysis aspect of the work to improve local air quality. For assistance in developing plans for the other key parts of the work, such as partnership building, education, communication, and mobilizing the community to take action, you will have to rely on members of your Partnership and on sources of assistance listed in Appendices A and B.

What if my community is interested in just one or a small number of facilities and their potential impact on our community?

The approaches discussed in this section and in the risk-based screening chapters of the Manual are designed for communities that have decided to review all outdoor sources of air pollution and to identify priorities for improving local air quality. Your community may choose to focus on a particular facility or source and its effects on the community or a particular location in the community. Analyzing the potential effects of releases from one or a small number of sources will take less time and resources. This effort could be used as a pilot to give the community a feel for what it can accomplish using a partnership. Of course, if you narrow the focus of your analysis of local air quality, you will not have enough information to identify which sources have the greatest impact on your community. So, if you do start with a narrow focus, we recommend continuing your work until you are able to include all sources of air pollution. If you choose to start with a focus on a particular source to begin your efforts, you will need to modify the options below and the chapters on risk-based screening to fit your focus.

What are the technical aspects of a work plan to understand and improve air quality?

The technical aspects of a work plan to understand and improve outdoor air quality can be broken into three parts:

1. A plan to identify the sources of toxics in community air and to collect information on the amounts and types of their releases, and information, if available, on air concentrations measured in or near the community

2. A plan to analyze all the information on releases and measured concentrations to figure out which chemicals and sources to identify as community priorities

3. A plan to identify ways to reduce risks from the chemicals and sources identified as community priorities

This section will describe options for addressing the first two technical parts of a work plan for improving air quality. Please see Figure 2-1 for an outline of the options presented here. For a discussion of the third part of a work plan, identifying ways to reduce risks for the sources and releases identified as community priorities, please see Chapter 8.

What are the options to consider to identify sources and collect information on their releases?

There are two approaches that the Partnership can take to accomplish this task.

Option One: The Partnership can use available state, tribal, and national databases to find information about the sources and releases affecting local air quality. Information on concentrations measured in or near your community may also be available in state, tribal, or national databases. Please see Chapter 3 of this Manual for a description of these databases and their contents.

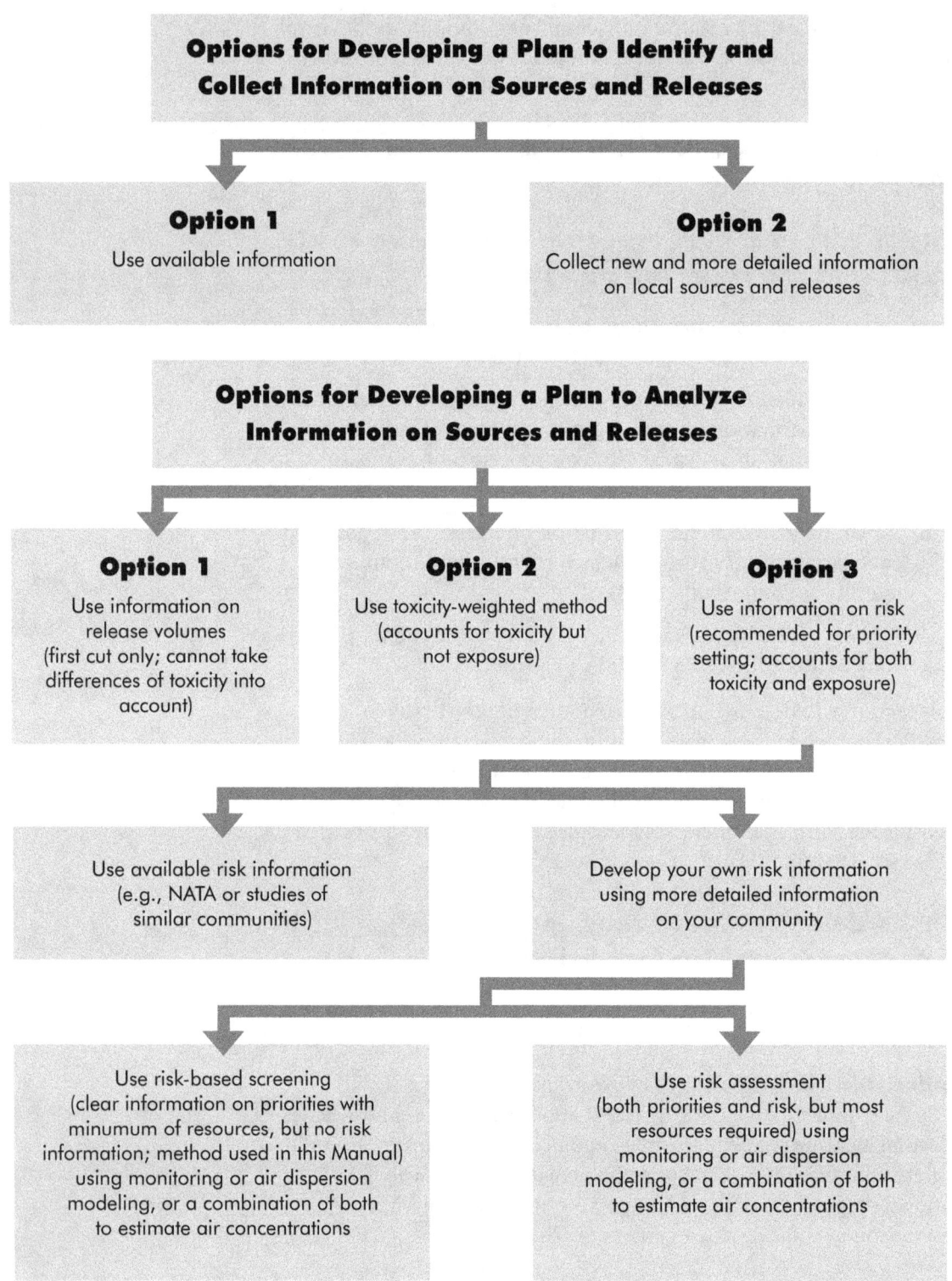

Figure 2-1.
Outline of Options for Developing a Technical Analysis Work Plan

Option Two: The Partnership can decide to collect new information to add to the information contained in the existing state, tribal, and national databases. This work to improve the information on sources affecting local air quality may be the most important work you can do to get a better understanding of local air quality. State, tribal, and national databases usually cover too large an area to include all the local details that may affect your air quality. For example, state, tribal, and national databases may not have the most accurate location information for some of the sources in your area, or they may not include some very small sources. Sometimes state, tribal, and national databases combine small sources and give only a county-wide total for the releases from these sources. Collecting detailed and accurate information on all large and small sources and their releases in your area will provide you with the best foundation for understanding local air quality and for identifying priorities for improving local air quality. Chapter 4 provides detailed guidance for collecting information on local air sources if you decide to include this as part of your work plan. Chapter 4 will also give you an idea of the resources you will need to collect this information.

In addition to planning for the collection of the information on sources and their releases, you will also need to decide on the method you will use to analyze this information to identify your priorities. Methods for analyzing information on releases will be discussed next.

What are options the Partnership can choose from to analyze the information on sources and releases to identify priorities for potential action?

The following are three options to consider to help you develop your plan for analyzing the information on sources and releases to identify priorities:

Option One: Comparison of release volumes: In this approach to setting priorities, a higher volume means a higher concern. This is the simplest method to use, and it can provide information to help set priorities, but its use as a priority-setting tool is very limited. This method should only be used as a first cut, since the toxicity of chemicals varies and the volumes of releases by themselves are not adequate to identify the chemicals that have the greatest impact on the community. For example, if this method is used, a highly toxic chemical released in small amounts might not be identified as a priority even though its impact on the community may be greater than a less toxic chemical with larger releases.

Approaches using

risk information

are the most

thorough

methods to use

to identify

priorities for

potential

community

action

Option Two: Comparison of toxicity-weighted scores calculated for releases: This method accounts for the varying toxicity of chemicals by assigning a value to each chemical to reflect the toxicity of the chemical: the more toxic the chemical, the higher the value. This value is usually multiplied by the release amount to produce a toxicity-weighted score for each chemical release. This approach can give a community a pretty good idea of the potential of releases to affect the community, and the toxicity-weighted scores are relatively easy to calculate. But this approach cannot account for differences in exposure to chemicals. For example, if the releases of a chemical are distant from the community, they may have little impact despite their toxicity or volume. And small releases of a toxic chemical close to a house or school would not be identified as a priority using this method even though they may affect the community. EPA's Risk-Screening Environmental Indicators (RSEI), a risk screening tool that is available to communities, includes a set of toxicity values that the Partnership could use for implementing this hazard-based approach. Information on the RSEI model and its toxicity values can be found at http://www.epa.gov/oppt/rsei/.

Other uses for the risk-related perspective provided by the RSEI tool are discussed below.

Option Three: Comparison of the risks of releases: The approaches using risk information are the most thorough methods to use to identify priorities for potential community action. If your community has sufficient resources, **this is the recommended option to use to identify priorities to improve air quality.** The potential impact of a release on a community depends on both the toxicity of the chemical and on the amount of exposure to the chemical that individuals in the community receive. Using risk information allows for a consideration of both of these elements. Because the risk approach combines information on exposure and toxicity, the impacts of any release amount and any release location, close to or distant from the community, can be more accurately portrayed. This enables a community to target its resources and energy to the reduction efforts that will have the greatest benefit for the community. Clear information on priorities also provides a firm foundation for building the consensus needed for effective action.

If a community decides to use risk information to analyze sources and releases, what approaches can it take?

If your Partnership decides to use information on risk to identify community priorities, there are two different approaches to choose from or combine to find the approach that best matches your goals and resources. First, you can use existing risk information on air releases for your community and communities similar to yours, and second, you can develop your own risk information for your community to add to and improve the existing risk information.

If you choose to develop your own risk information, you can use a risk-based screening approach, the approach described in this Manual, or you can conduct a risk assessment. Each of these options is discussed briefly below.

How could a community use existing risk information to set community priorities?

A good deal of information on the risks from toxics in ambient air is now available to communities. As a part of EPA's National Air Toxics Assessment (NATA), EPA has conducted a national-scale assessment of 33 air pollutants (a subset of 32 air toxics on the Clean Air Act's list of 188 air toxics, plus diesel particulate matter, or diesel PM). This assessment, which will be updated to reflect new release information every three years, when used in combination with other local-scale information, is a valuable source of risk information for communities. In addition to EPA's national-scale assessment, detailed studies of air toxics risks for a major metropolitan area and for several urban neighborhoods have also been completed, and the results of these studies are available. They can be used to give you an understanding of the risks in areas that may be similar to yours. Studies of neighborhoods in Baltimore, St. Louis, Port Neches and Houston, Cleveland, and Portland are available. Other studies of air risks in neighborhoods and cities were under way at the time of publication of this Manual and may now be available. A clearinghouse for studies of ambient air risks, including those mentioned above, can be found at http://epa.gov/ttn/atw/urban/urbanpg.html.

Additional valuable information on urban risks can be found in the study of ambient air in the Los Angeles metropolitan area conducted by the California's South Coast Air Management Board. This study can be found at http://www.aqmd.gov.

In addition to studies of neighborhoods and cities similar to yours, information that allows you to easily perform screening-level analyses of chemical releases from most of the largest facilities in your area is available from EPA. The RSEI model uses Toxics Release Inventory (TRI) data to perform screening-level analyses comparing chemical releases using risk-related as well as hazard-related and pounds-based perspectives. This model is easy to access and run, so you will be able to use this tool to help identify priorities for your community. Information on accessing and using RSEI can be found at http://www.epa.gov/oppt/rsei/.

It is probably a good idea for every partnership to start by collecting information that is readily available on releases and risks

Please remember that this tool, since it only compares releases from large facilities, needs to be used with information on other sources of air toxics, such as mobile sources, to adequately identify community priorities.

You may want to assign a Partnership team to gather and summarize the available information on both releases and risks from air toxics and present this information to the Partnership. This effort could take several weeks, so if possible, it would be good to get a team started on this work while the Partnership works to clarify its goals. When all the available information on releases and risks relevant to your community is assembled, you may then decide that you already have enough information on risks from air toxics in communities similar to yours for you to identify the priorities that you will work on in your community. For example, in the review of existing information, you will find that risks from some sources, such as mobile sources, are consistently high in all the available studies. Based on this information, you may decide that there is sufficient information to target these sources as priorities. Several neighborhoods in Cleveland took this approach, enabling them to focus quickly on the work to find solutions. A summary of the Cleveland project can be found at http://www.ohiolung.org/ccacc.htm.

It is probably a good idea for every partnership working on improving air quality to start with this step of collecting the information that is already available on releases and risks. Even if you decide that you need to do more study of the risks in your community, this information will be very informative, and it may point to some priorities that you can begin to work on immediately.

As an option, you may consider collecting the available release and risk information described above and see if you can use this information to reach agreement on some obvious priorities to begin work on immediately, and, at the same time, organize the Partnership to collect the information for a detailed local inventory of sources and releases. When this inventory is completed, you can then review the information and decide whether your community will need to do additional work to use risk information to analyze the new information on releases and sources to see if there are additional sources and chemicals to add to the community's list of priorities.

When should the Partnership consider developing new risk information to help set community priorities?

Some communities may decide that the available information on risk is not sufficient for them to identify community air toxic priorities, and they may decide to develop their own risk-based analysis of the toxic releases affecting their community. There are several reasons a community might choose to do this analysis. In some communities, a partnership may find that it needs to do its own analysis to come to an agreement on priorities. Some members of a partnership may consider the situation in their community to be different from other communities, or they may think that available studies do not accurately portray the releases in their community. Some communities may also feel that the limited number of chemicals covered in available risk studies may not be sufficient to understand the risks in their community. Working together on an analysis of community air risks can help give the participants a sense of ownership of the process and a willingness to commit to decisions based on the analysis.

Working together on an analysis of community air risks can help give participants a sense of ownership of the process and the results

Another reason a partnership may consider developing its own risk information may be to analyze releases from small businesses more carefully. The National Air Toxics Assessment (NATA), because it is a national-scale study, sometimes combines releases from small businesses to estimate risks. Communities may also feel more certain of the accuracy of risk information if they deal directly with facilities and ensure that the release information used in the analysis is accurate and up to date. And, if a community has identified the strengthening of community capacity as a goal, conducting a community risk-based analysis may provide an opportunity to educate the community on the details of risk science and air quality. Some communities may also want to learn how to develop their own risk information so that they will have the ability to use this capacity to provide risk information on any changes that may occur if new facilities or releases are being considered in their communities.

If a community decides to develop new risk information to set priorities, what are the approaches it can take?

If you decide to do your own risk-based analysis to supplement existing risk information, you will have two basic approaches to use: risk-based screening and risk assessment.

Risk-based screening is designed to identify the chemicals and sources that have the greatest potential to affect the health of some or all community members, using the minimum possible expenditure of community resources. The method of analysis described in Chapters 3 through 12 of this Manual is an example of this approach. Risk-based screening minimizes resources by screening out low-risk chemicals and by simplifying the estimation of exposure to develop relative risk estimates for each chemical concentration and each source. Because of these simplifications, risk-based screening has limitations. It can identify chemicals above screening levels and identify the chemicals and sources with the greatest potential impact on the health of members of the community. In most cases, this gives communities the information they need to set effective priorities. But risk-based screening does not estimate the actual risks resulting from releases and sources. And, because risk-based screening does not attempt to estimate risk, it cannot be used to estimate the cumulative risk from all or some of the chemicals. To get a more detailed overview of the risk-based screening method, its advantages and limitations, and an idea of the resources required to use this method, please see the introduction to risk-based screening in Chapter 3.

Risk assessment will be necessary if your community decides that it needs or wants to estimate the risk from chemicals and sources or the cumulative risk from multiple chemicals and sources. Information on risk and cumulative risk may be important to reaching a consensus on action. Information on risks from air toxics will also allow you to compare these risks to other community risks so that you can set priorities to reduce risk most effectively. Risk assessment is also a good indicator that can be used to measure progress. Estimating risk goes beyond risk-based screening by replacing the simplified method of estimating exposure used in risk-based screening with an additional step to more accurately estimate actual exposures in the community. As a result, risk assessment requires more resources and more technical expertise than risk-based screening. For detailed guidance on estimating risks at the local level, please see EPA's Air Toxics Risk Assessment Library. This resource can be found on the Internet at EPA's Fate, Exposure, Risk Assessment (FERA) site at http://www.epa.gov/ttn/fera/.

As an option, a partnership may decide to combine parts of the risk-based screening method outlined in this Manual with risk assessment. For example, a partnership could use the screening method to identify the higher-risk chemicals and sources and then conduct a risk assessment on only those higher-risk chemicals. A screening step is, in fact, a part of most risk assessments, including both the national-scale assessment done as a part of the NATA and the study done of the Los Angeles metropolitan area.

In the future, new tools and models will be available to help communities with risk assessment. These tools may reduce the resources needed for risk assessment, making it a practical option for most communities. Tools now under development or modification to make them accessible for community risk assessments include E-FAST, a screening-level model available now on the Internet, and IGEMS, a model that will combine air dispersion and exposure models with toxicity information to estimate risk. IGEMS will also include geographic information system (GIS) capability as a future enhancement. A description of E-FAST and IGEMS and updates on their development can be found at http://www.epa.gov/oppt/exposure/.

Other tools for the assessment of air pollutant fate, exposure, and risk are available on EPA's Technology Transfer Network web page at http://www.epa.gov/ttn/.

Included among these tools are HAPEM, the model used for estimating exposures for EPA's National Air Toxics Assessment as well as other exposure models. EPA's Region 6 is also developing a risk-based tool, RAIMI (Regional Air Impact Modeling Initiative), that will characterize risk-related impacts and include GIS capability. Information on this model and its availability can be found at http://www.epa.gov/Arkansas/6pd/rcra_c/raimi/raimi.htm.

Are there different approaches to risk-based screening and risk assessment for the Partnership to consider?

There are different approaches that risk-based screening and risk assessment can take. Approaches differ primarily in the method used to determine the concentrations in community air. Air concentrations can be obtained in two ways: by actually measuring the concentrations (monitoring) or by using air dispersion modeling to estimate the concentrations. Risk-based screening and risk assessment can be accomplished using one or a combination of these two methods for developing air concentrations.

Monitoring directly measures what is in the air either at fixed locations in a community or as the air is breathed by someone living in the community through personal monitors worn by community volunteers. Since monitoring relies on neither estimates based on computer modeling nor on the reliability of release data, it can provide accurate measures of concentrations at specific locations or for specific individuals. Setting up new monitoring is an appropriate and thorough way to analyze community air, but it can be expensive and it will require at least a year of measurements for estimates of long-term effects. Because of its cost, monitoring can usually be done in only a limited number of locations in a community. The information on concentrations from monitoring also has some limits. By itself, monitoring does not provide information on the sources of the chemicals measured at the monitoring location. And most monitoring involves sampling air at regular intervals, so monitored concentrations can represent only the concentrations in the air at the times of measurement. And, because of the expense or the availability of appropriate analytical methods, the list of chemicals measured in monitoring will also be limited.

Air dispersion modeling is also a valid tool used by regulatory agencies and health agencies to estimate concentrations in community air. Air dispersion modeling also has some advantages because it does allow a community to estimate concentrations at any location in the community and it does identify the sources of the concentrations. Risk-based screening studies that use air dispersion modeling can differ depending on the number of sources that are modeled. Accurate air dispersion modeling of all releases provides the most information, but collecting and using detailed information on a large number of sources can be costly and resource intensive. To conserve resources, tiered screening approaches, such as the one described in Chapters 3 through 12 of this Manual, can be used to limit the amount of air dispersion modeling that is required to make decisions.

Communities can use monitoring and modeling in different amounts depending on their resources and goals. If a community has resources for either monitoring or comprehensive modeling, it will be able to get more accurate or complete information. Complete monitoring and comprehensive modeling can, for example, be used as part of a risk assessment to answer questions about the cumulative risk resulting from all chemicals in community air. Tiered screening methods, such as the one described in this Manual, cannot be used to address cumulative risk from all chemicals.

Community and government partnerships are currently engaged in assessments using different combinations of modeling and monitoring.

Community residents inspect a neighborhood air monitoring station

- For an example of a community study based on monitoring see St. Louis Community Air Project at http://www.stlcap.tripod.com/.

- For an example of comprehensive modeling, see EPA Region 6's Port Neches study at http://www.epa.gov/Arkansas/6pd/rcra_c/raimi/raimi.htm.

- For an example of combined modeling and monitoring, see the South Coast Air Management District's study of air in the Los Angeles regions at http://www.aqmd.gov.

The risk-based screening method described in Chapters 3 through 12 of this Manual is designed to use the minimum of resources to identify priorities. It uses monitoring information if it is available but it does not require new monitoring. Instead of new monitoring, it relies on air dispersion modeling to estimate concentrations, and it uses a tiered screening process to limit the amount of air dispersion modeling needed to make decisions. For a more detailed description of this method, please see Chapter 3.

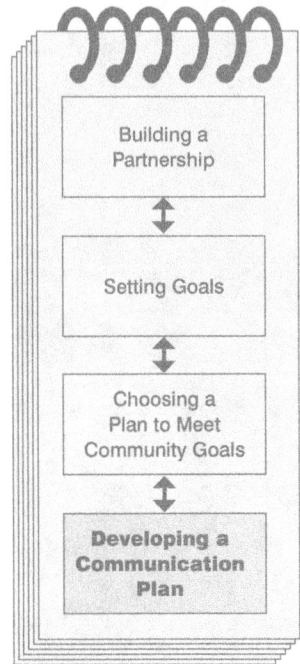

Getting Started

Building a Partnership

↕

Setting Goals

↕

Choosing a Plan to Meet Community Goals

↕

Developing a Communication Plan

Developing a Communication Plan

Why should the Partnership focus on communication?

The success of the Partnership depends on the participation of the community for setting goals and standards, for collecting information, and for mobilizing the resources that will be needed to improve air quality. Participation of the community will depend on the Partnership's ability to answer community questions about air quality and to develop a consensus in the community on priorities and on a plan to improve air quality. As a result, providing information to and getting input from the community are at the heart of the Partnership's work.

A list of resources to support the communication work of the Partnership can be found in Appendix A. This section of the Manual addresses communication issues particular to air quality partnerships.

How much time and effort will it take for the Partnership to communicate effectively?

Because the participation of the broad community is crucial to the success of the Partnership, it is likely that communication will take as much time and effort as the collection and analysis of information. Time will be needed to develop effective communication materials, hold meetings in the community, and revise Partnership plans based on community input. Partnerships may

Partnership team presents its work at a community meeting for discussion.

want to consider the recommendation of the Baltimore Partnership that the focus of meetings of the core Partnership group should alternate between conducting the screening analysis and planning outreach to the community. This schedule allows for the time the Partnership group will need to develop and provide regular updates for the community.

When should the Partnership communicate with the community and its stakeholders?

Beginning with the outreach to develop and distribute its goal statement, the Partnership should communicate with the community regularly throughout the course of its work. This interaction will ensure that the community is informed and current about Partnership activities and has adequate opportunity to provide input into the Partnership's direction and activities. The process of communicating the information developed by the Partnership and developing the consensus in the community that will be needed to improve air quality will take time. Regularly scheduled, step-by-step communication of the work of the Partnership, and regular feedback from the community throughout the course of the Partnership's work, will provide the best chance for success.

Communication will take as much time and effort as the collection and analysis of information

Who in the Partnership will work on communication?

Although there may be several members of the core Partnership group that take primary responsibility for preparing communication materials and organizing outreach, all members of the core group are likely to participate in the communication efforts. Each member will take the initiative to communicate directly with the organizations or groups that they represent. For example, industry representatives could take the initiative to keep the business community informed and involved, and representatives of community churches could do the same for their own organizations. The core Partnership group may also want to organize teams to plan for once-a-month meetings with community groups, organizations, or block clubs to explain the Partnership's progress and to gather input from the community. Different members of the core group will also be called on to explain aspects of the work to large community meetings and to local media. Effective communication will require a joint effort from all the members of the Partnership's core committee.

Who should the Partnership include on its list for regular communication?

The goal of the Partnership should be to reach as many members of the community as possible. Targeting existing community organizations and institutions may be the most effective way to communicate with the community. The Partnership can meet regularly with key organizations, such as community and neighborhood associations, churches, schools, parent-teacher associations, community clubs, youth organizations, housing associations, and business associations. Reaching out to these groups regularly with information and encouraging and assisting them to communicate this information to their members will be important. Local media and local libraries will also be important for communicating information to the community.

In addition to outreach to the community in general, the Partnership should also consider targeting community leaders to keep them informed about the work of the Partnership. A list of elected officials and leaders of community organizations can be developed and plans made for interacting regularly with these leaders.

It will also be important for the Partnership to develop a list of key stakeholders outside the community that will need to be informed about the work of the Partnership. Local, state, tribal, and national elected officials and organizations that may be interested in the work of the Partnership could be included on this list and updated on a regular basis.

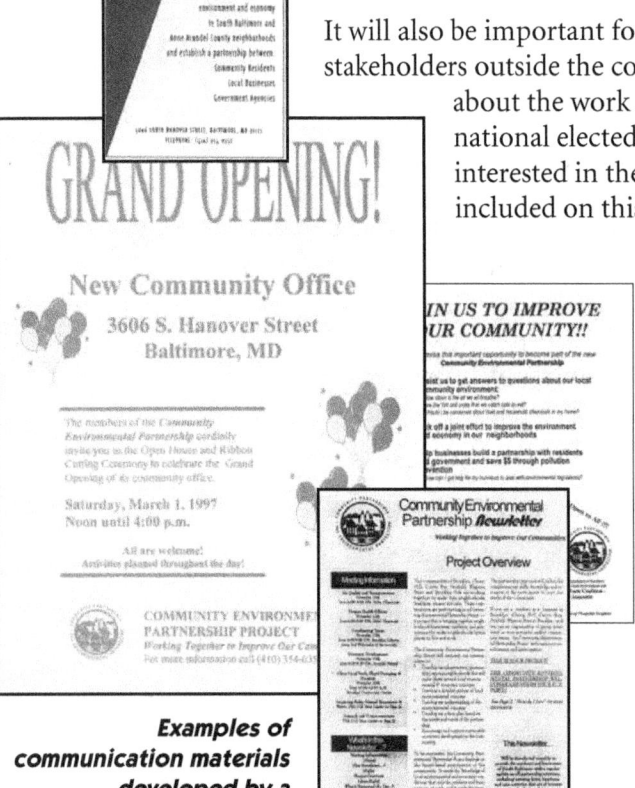

Examples of communication materials developed by a Partnership

What methods could the Partnership use to prepare understandable communication materials?

Presenting information to community groups and paying attention to their questions and responses are good ways to learn how to develop materials that can communicate effectively. With regular practice, Partnership communication should improve over the course of the project. Involving members of the community who are not working directly in the Partnership to help prepare Partnership materials can also help. Local teachers and high school classes can also be a resource for preparing effective communication materials. For key Partnership reports and information, such as the goal statement and the final report, it may be helpful to try out materials in small groups in the community before finalizing them for wide distribution.

Should the Partnership develop a communication plan?

A communication plan is a description of the activities that the Partnership will undertake to communicate with the community and stakeholders. Developing a written communication plan at the start of the project will help the Partnership to plan adequately for this work.

A communication plan could include some or all of the items listed on the following page.

What are the next steps once the Partnership has been built, goals have been clarified, and plans for work and for communicating with the community have been developed?

The following chapters of this How-To Manual are designed to help a community use risk-based screening to understand and improve air quality. Step-by-step guidance is provided for all aspects of risk-based screening.

If your partnership has chosen a different approach to improving air quality, then some of the content of the following chapters will not be relevant to your work. But even if you choose a different approach, some of the information may still be useful. For example, many communities will be interested in collecting information on the sources and releases in their areas. Guidance for collecting this information and building a local source inventory can be found in Chapter 4. Communities can also find some help in identifying solutions to their concerns in Chapter 8 and a description of air dispersion modeling and its use in Chapter 6. Please feel free to use any part of this Manual if it can help you to understand and improve your air quality.

Developing a written communication plan at the start of the project will help the Partnership to plan adequately for this work

Communication Plan Checklist

✔ A discussion of how the core Partnership group will organize itself to carry out the communication work: This could include the identification of who will lead and participate in the communication work and who will have the main responsibility for communicating with each sector of the community and with each of the organizations and governments participating in the Partnership.

✔ A list of the key milestones in the Partnership's work that will require a communication effort: This list could include the completion of the goal statement and each of the five other steps of the screening process. Partnership initiatives requiring community volunteers will also require communication efforts.

✔ A plan for facilitating community input and response to Partnership communications and activities: This could include large and small community meetings, comments in writing or responses over the Internet, focus groups and telephone surveys.

✔ A description of the kinds of outreach materials the Partnership will produce, such as newsletters, reports, presentations, talking points, posters, leaflets, brochures, press releases, an Internet web page, e-mails, and letters.

✔ A description of how these outreach materials will be written and copied or printed for distribution.

✔ A plan for distributing Partnership communication materials in the community, including where and how to distribute materials.

✔ A list of contacts for local press and media and a plan for communicating with and providing Partnership materials to them on a regular basis.

✔ A list of key local organizations that the Partnership will rely on to distribute outreach materials to their members, and a plan for meeting regularly and providing speakers, outreach materials, and assistance to these organizations.

✔ A list of key community leaders and key stakeholders outside the community that the Partnership will need to keep informed and a plan for communicating regularly with them.

✔ A plan for communicating Partnership information to the organizations and governments participating in the Partnership.

✔ A plan for handling communication for any crisis or emergency that may occur during the Partnership's work. For example, if an accidental chemical release should occur in the Partnership area during the course of the work, the community will look to the Partnership for assistance. Developing a plan for communicating in this and similar situations would help prepare the Partnership to respond effectively.

An Introduction to Risk-Based Screening

What is risk-based screening?

Risk-based screening is a method a community partnership can use to identify all the chemicals in the community's outdoor air, estimate the concentrations of those chemicals in community air that result from the combined releases from all sources, and analyze this information to find the chemicals and sources that have the greatest potential to affect the health of some or all community members. Communities using risk-based screening choose a level of risk and use this level to screen chemicals and sources to find those that present a risk above the screening level to some or all community members. This information allows a community to set priorities and focus its efforts to improve air quality where they will do the most good.

Organize
Collect
Analyze
Mobilize

The risk-based screening method described in this Manual is also a process that will help to develop the agreement you need in your community to get things done. Doing the work will also build the long-term capacity of your community to understand and address air quality concerns that might come up in the future.

Is there an example of a community using risk-based screening that can help us understand how it would work?

To help you understand how this might work in your community, here is an example of how this process worked in the industrial neighborhoods in Baltimore, where the screening process described in this Manual was first developed. For years, residents of several industrial neighborhoods in Baltimore had concerns about the potential impacts of chemical releases from all the many sources in and around their community. To try to address these concerns,

Completing the screening process allowed the Baltimore Partnership to set priorities and focus its energies where they could most effectively improve local air quality

community organizations, governments, and a local university formed a partnership to work together to improve local air quality. Through this partnership, the knowledge of the details of the local community and the scientific expertise needed to understand the potential impacts of chemical releases on community health were combined. The partnership began by collecting information from government databases and from local residents on all the releases and sources in and around the partnership area. The Baltimore partnership found 172 different chemicals in community air coming from 125 different sources. The Partnership then estimated the concentrations in community air that resulted from these releases and used the risk-based screening process to sort through these estimated concentrations to find the community's priority chemicals and sources. The Baltimore Partnership found four chemicals released from 17 different sources with concentrations in community air that exceeded the community's risk screening level. These chemicals and sources were identified as community priorities, and teams were formed to develop recommendations to address the chemicals and their sources.

Completing this screening process to identify the chemicals and sources that presented the greatest potential to impact the health of members of the community allowed the Baltimore Partnership to set priorities and focus its energies where they could most effectively improve local air quality. A Baltimore Health Department official summed up the value of the screening project like this: "For many years we have all been trying unsuccessfully to answer the concerns of these neighborhoods about their air quality. With the partnership and the screening process, we now, for the first time, have the ability to work together to begin to address those concerns."

What questions can communities answer using risk-based screening?

The Manual will help communities answer questions such as:

- What are the sources of outdoor air pollution in my community?

- What are the levels of chemicals in community air that result from the combination of all of the sources in and around our community?

- Are the levels of chemicals in community air high enough to be identified as community priorities to evaluate for possible action?

- What are the sources of the chemicals that are identified as community priorities?

- What should our priorities be in working to improve air quality?

- How does the air in our community compare with other communities?

What results could a community expect from using the risk-based screening process described in this Manual?

The following results can be achieved by carrying out the science-based process described in this Manual:

- An **inventory** of all significant sources of pollutants in outdoor community air, with information about type and quantity of chemicals emitted to the air in the study area

- Estimates of the **concentrations of chemicals** in community air that result from all the sources in and around the community

- **A comparison of chemical concentrations in community air to risk-based screening-level benchmarks** set by the community to identify priorities to be evaluated for possible community action

- Clear **priorities** for focusing community efforts for actions on the chemicals and sources that present the greatest potential to impact the health of members of the community

- A **baseline** and the ability to measure progress in improving air quality

- An **increased community capacity** to understand and address air issues in the long term that results from the knowledge, understanding, and trust gained in completing the process

- A **better agreement within the community** on air issues based on the improved understanding provided by the work

- The **ability to compare** community air quality to air quality in other communities where air concentrations have been measured or estimated

As discussed in Chapter 2, the screening process described in this Manual is designed to identify priorities based on health risks due to the chronic, long-term health effects of chemicals in outdoor air. Risks due to short-term, acute effects of chemicals in outdoor air and risks due to indoor air exposures are **not** considered by the method considered in this Manual. Please see a more detailed discussion of the limits of the method used in this Manual at the end of this chapter.

What's in the Manual that allows a community to get these results?

Chapters 3 through 12 of the Manual provide a step-by-step guide to using the environmental science and information needed to understand and improve local outdoor air quality. Science-based tools, such as risk-based screening, are available to help make sense of local environmental data, but not usually in a form accessible to communities. This Manual is an attempt to make this science accessible so that communities can identify and find solutions to their concerns. The Manual presents and explains the science of risk-based screening and the process that a community can use to identify and inventory local sources of chemical releases, to review these sources to identify known hazards that might present a health risk to the community, and to set priorities for action to improve local air quality if they are needed.

Combining the local knowledge of the community and its businesses with the science of air quality specialists is an effective way to answer questions about local air quality

In addition to making science accessible to communities, the Manual incorporates a process designed to bring all sectors of the community and scientists together to share information and to work to improve local air quality. Combining the local knowledge of the community and its businesses with the science of air quality specialists is an effective way to answer questions about local air quality and to mobilize the resources needed to make improvements. In the process described in this Manual, non-technical community residents work side by side with technical experts to carry out all aspects of the effort, with each contributing essential parts to the analysis and work. The process, and the sharing of information that takes place to complete the process, helps to build the trust and the agreement needed to take effective action to improve air quality.

Is risk-based screening more suitable to support voluntary or regulatory actions?

Risk-based screening is designed to make risk science accessible to communities with limited resources. It uses enough science to allow a community to identify priorities and develop an effective plan for action. Risk-based screening is not a full risk assessment and, as a result, it is most appropriate for use by community partnerships committed to working together to find voluntary solutions to community air concerns. Non-voluntary regulatory or enforcement actions are likely to require the additional analysis of a full risk assessment. The work done to complete a risk-based screening project can, if necessary, serve as the foundation for a full risk assessment.

Who is the intended audience for Chapters 3 through 12 of this Manual?

Chapters 3 through 12 are designed for community partnerships interested in using risk-based screening to understand and improve local outdoor air quality. The Manual is designed for both the non-technical and the technical members of a community partnership effort. The Manual can also be used by community leaders familiar with air quality issues to get an overview of the risk-based screening method to decide if it is an appropriate approach for meeting community goals.

How is the description of risk-based screening in Chapters 3 through 12 organized and how should these chapters be read?

This Manual is divided into two basic parts: An **Overview** section, Chapters 3 through 8, that contains the information that everyone working in the Partnership will need to know to fully participate in an effort to use risk-based screening to understand and improve air quality, and a **Technical Guidance** section, Chapters 9 through 12, that contains the detailed information that the technical members of the Partnership will need to carry out the technical aspects of the risk-based screening methodology for the Partnership.

This chapter, Chapter 3, "An Introduction to Risk-Based Screening," provides a summary of the risk-based screening method used in this Manual as well as information on the resources that a partnership will need to implement the method described in the Manual. The Partnership can use this information on the resources that will be needed to carry out a risk-based screening project to begin planning and organizing its work. The remaining chapters of the Overview, Chapters 4 through 8, provide the basic information that the members of a partnership will need to carry out each step of the process.

If a community partnership decides to use this Manual, it may want to review all of these chapters at the beginning of a project and take them up again, one at a time, as the project progresses. Partnership members who read the Manual on their own should not be surprised or discouraged if they find some parts of the Manual to be unclear or

Each partnership will have to draw on its members throughout the process to provide additional explanations and illustrations to explain the concepts and methods used in the Manual

confusing at first. The chapters of the Manual are designed to cover all the elements of the screening process, but they are only summaries and they will need to be supplemented with further explanations by members of the Partnership who have expertise in the different areas of the screening and partnership process. Both technical and non-technical members of a partnership will find sections of the Manual that they may not understand completely without supplemental explanations. For example, the descriptions of scientific concepts like risk and of technical tools like air dispersion modeling are brief, and the non-technical members of the Partnership will probably need additional explanations and illustrations to ensure that they can fully participate in decisions involving these concepts. Similarly, community residents will have to explain and provide examples of local knowledge to the technical members of the Partnership to supplement the Manual. Each partnership will have to draw on its members throughout the process to provide the additional explanations and illustrations that will be needed to adequately explain the concepts and methods used in the Manual. In addition, references to additional support materials that could be used to help supplement this Manual are made throughout the text. If your Partnership develops an effective tool to explain or illustrate a part of the screening process, please be sure to share it with other communities.

Both the Overview section and the Technical Guidance section are organized around six steps, beginning with partnership building and ending with the implementation of recommendations to improve air quality. The six steps will be discussed later in this chapter.

How was this Manual developed?

This Manual was developed over several years in a multi-step process. As a result, it incorporates the ideas and experiences of a very large number of people. The risk-based air screening process described in this document was originally developed by a partnership for the study in Baltimore, Maryland, that is described above. In this effort, residents, local organizations, schools, industry, and government worked together for three years to design and implement an air screening methodology for five Baltimore neighborhoods. As mentioned above, these neighborhoods faced the challenge of trying, with limited resources, to understand the potential impact on their community of 125 facilities with 400 releases of 175 different chemicals. The result of this Partnership effort was a screening methodology that formed the starting point for this Manual. A summary of the work of the Baltimore Partnership is available in a case study titled *Baltimore Community Environmental Partnership Air Committee Technical Report, Community Risk-Based Air Screening: A Case Study in Baltimore, MD* (EPA 744-R-00-005). This document can be downloaded or ordered on the web at: http://www.epa.gov/oppt/cahp/case.html.

A summary of lessons learned by the Baltimore Partnership, a part of the case study, can be found in Appendix D. We have tried to incorporate these lessons into the guidance contained in this Manual.

Following the completion of the efforts in Baltimore, the screening methodology was reviewed by independent peer reviewers, EPA staff, and community leaders. Based on the suggestions provided by these reviewers, the methodology developed in Baltimore was expanded and improved to develop this How-To Manual.

Finally, the Manual was sent out again for both an internal EPA review and a formal external peer review by a panel of independent experts. Both of these reviews resulted in significant improvements in the Manual. Lists of both the EPA and the external independent reviewers of this Manual are provided on the acknowledgment page of this Manual. Information on how to get a summary of the external peer review comments and an explanation of how they were addressed in the development of this Manual can be found on the internet at: http://www.epa.gov/oppt/cahp/howto.html.

EPA hopes to amend and update the methodology and Manual as communities gain more experience in their efforts to understand and improve their air quality. Future updates to this Manual will be available on the web site provided above.

Should a community using this Manual view it as a finished product or make changes to the Manual as needed?

This Manual is a summation of experiences of the Partnership in Baltimore and the ideas of all those who helped with or commented on this effort. Although every effort was made to make it as useful as possible, the fact that it is based on limited experience means that it can be improved as other communities develop additional practice. Every community is different and will have to address issues specific to their community, so this Manual should be seen as a work in progress. As the EPA team that pulled together all the comments to produce this Manual, we encourage you to share your experiences with us so that the Manual can be updated and other communities can benefit from your experiences. You may visit the How-To Manual web site to learn how to share your experiences or look for recent updates to the Manual. The Manual web site can be found at: http://www.epa.gov/oppt/cahp/howto.html.

How does risk-based screening described in this Manual work to help a community identify priorities?

Risk-based screening combines information on concentration and toxicity into a practical method that allows a community to identify priorities with the minimum possible use of resources

To identify community priorities, the How-To Manual first explains how a community can estimate the concentration of each of the chemicals in community air that result from all the sources of that chemical in and around the Partnership area. Using concentrations to set community priorities is much better than just looking at the volume of releases because the concentration, not the volume released, is the key to understanding the effects that the chemical might have on individuals in the community. Small releases can result in high concentrations in parts of the community that are close enough to the source. On the other hand, large releases may result in only small concentrations in the community if the source is far away or if the releases come from a tall stack. (The details of how to estimate the concentrations are discussed below and in the subsequent chapters of the Manual.)

But just estimating concentrations alone would not be sufficient to tell communities where to focus their energies. The Manual also allows the community to incorporate information on the toxicity of the chemicals released into its air. A low concentration of a very toxic chemical may be more harmful than a high concentration of a less toxic chemical. The How-To Manual allows communities to combine information on the concentration and the toxicity of the chemicals in their air to decide which chemicals should be targeted for potential community action. (The details for using information on the toxicity of chemicals can be found in the "Initial Screen," Chapter 5.)

Finding a practical way to develop and use information on both the concentrations and the toxicity of chemicals was the challenge faced by the Partnership in Baltimore. The Partnership chose risk-based screening as the method it would use to identify the chemicals it needed to target. Risk-based screening combines information on concentration and toxicity into a practical method that allows a community to identify community priorities with the minimum possible use of resources. This Manual is designed to make the risk-based screening method developed in Baltimore more accessible to communities.

Risk-based screening is widely used by scientists to identify contaminants of concern. It works by setting an acceptable level of risk and then determining the concentration of a chemical that would result in that level of risk in a hypothetical individual exposed to the concentration, usually an individual with a reasonable maximum exposure. This concentration is set as the risk-based screening concentration for that chemical. (An alternate approach to risk-based screening that uses the proportion of total risk to screen chemicals instead of screening-level concentrations is possible. Although this Manual does not provide detailed guidance for that approach, a fuller explanation of screening by proportion of risk can be found in Chapter 5.)

To adapt risk-based screening for community purposes, a community first decides on the level of risk that it finds adequate to protect community health. Based on this decision, the community calculates the concentration that would result in the chosen level of risk in a hypothetical community member who is exposed to that concentration. This concentration becomes the community risk screening level. Deciding on how the risk screening concentrations are set allows the community to choose levels that match its values and needs. If a chemical concentration in community air is found to be higher than the screening concentration, then that chemical will be identified as a priority to be evaluated for possible community action. Chemicals with concentrations that are lower than the community screening concentrations can be set aside as lower priorities. (Chemicals set aside should be reevaluated periodically to account for changes in information.) As a result, by using the risk-based screening method, the community will be able to identify all the chemicals that need its attention. This will allow the community to focus its limited resources in a way that will most effectively improve the community's health and environment.

Using risk-based screening to improve local air quality depends on an agreement in the community to work together to try to find ways to reduce releases of chemicals that have concentrations above the risk screening level. Reaching this agreement, and developing the understanding of local air quality that it requires, are key to the success of the method described in this Manual. (A discussion of the task of setting the community's screening level of risk can be found in the "Initial Screen," Chapter 5.)

Using risk-based screening to improve air quality depends on an agreement in the community to work together to try to find ways to reduce releases of chemicals that have concentrations above the risk screening level

Here is a list of the technical tasks described in the Manual that a community partnership will need to complete to identify priorities for potential action to improve its air quality:

- Identify and collect release information on all sources of air pollution in and around the Partnership area.

- For each chemical released, estimate the air concentration in the community that results from all the sources of the chemical.

- Choose a level of risk to serve as the community's risk screening level.

- Calculate the risk screening concentration, that is, the concentration equivalent to the risk screening level, for each chemical in community air.

- Compare the concentration for each chemical in community air to its risk screening concentration and identify chemicals above their screening level as community priorities.

Once the community has identified priority chemicals, what actions can be taken to reduce their levels?

Finding and implementing effective ways to improve local air quality are the goals of the effort described in this Manual

Finding and implementing effective ways to improve local air quality are the goals of the effort described in this Manual. The screening analysis described in the Manual is designed to help all sectors of a community work together to get a better understanding of local air quality. This common understanding will form a solid foundation for taking action, if necessary, to improve local air quality. This action, discussed more fully in Chapter 8, could take various forms, such as a voluntary effort by local businesses to reduce the emissions of priority chemicals, work to address mobile source emissions, or a campaign to reduce the use of toxic chemicals around the home. Communities might also decide to work with public health agencies to identify and address community health concerns. Monitoring or further analysis to verify or clarify issues raised in the screening analysis may also be necessary. Finding and taking effective actions to improve the community's health and environment are the goals of the work described in this Manual.

What does the Manual do to help limit the resources a community will need to use the risk-based screening method?

Completing all of the analysis described above will require significant community resources. Estimating the concentrations in the air that result from all the releases, the second step described above can, in particular, require significant technical resources. Because many communities have limited technical resources for estimating air concentrations, this Manual uses a tiered screening process designed to minimize the resources needed to estimate concentrations. The first tier of the screening process estimates concentrations using a simple look-up table that allows the Partnership to screen out chemicals with a minimum of effort. The second and third screening steps repeat the analysis, each time using better information and more sophisticated methods to estimate concentrations. The use of this tiered approach means that the Partnership will not have to use the resource-intensive effort needed to accurately estimate concentrations for all the chemicals in community air. Each of these screening steps is described in more detail below and in Chapters 5, 6, and 7.

What steps would a partnership take if it uses this Manual?

The overall methodology described in this Manual, including partnership building, the tiered risk-based screening approach, and the work to improve air quality, can be carried out in the six steps outlined below. Each of these steps is described in more detail in a separate chapter of the Manual. The chapters detailing each step are listed below. Please see Figure 3-1 for a summary of the community air screening methodology and Figure 3-2 for a detailed description of each of the six steps for the air screening methodology discussed in the Manual.

Step 1: Build a Partnership, Clarify Goals, and Develop a Communication Plan (Chapter 2)

The process begins with the effort to form a broad partnership to understand and improve local air quality. The Partnership should aim to include as many sectors of the community as possible. Interested local citizens, community leaders, and representatives of community organizations, the local business community, schools, churches, nearby colleges and universities, and local, state, tribal, and federal governments should all be included, if possible. The Partnership will

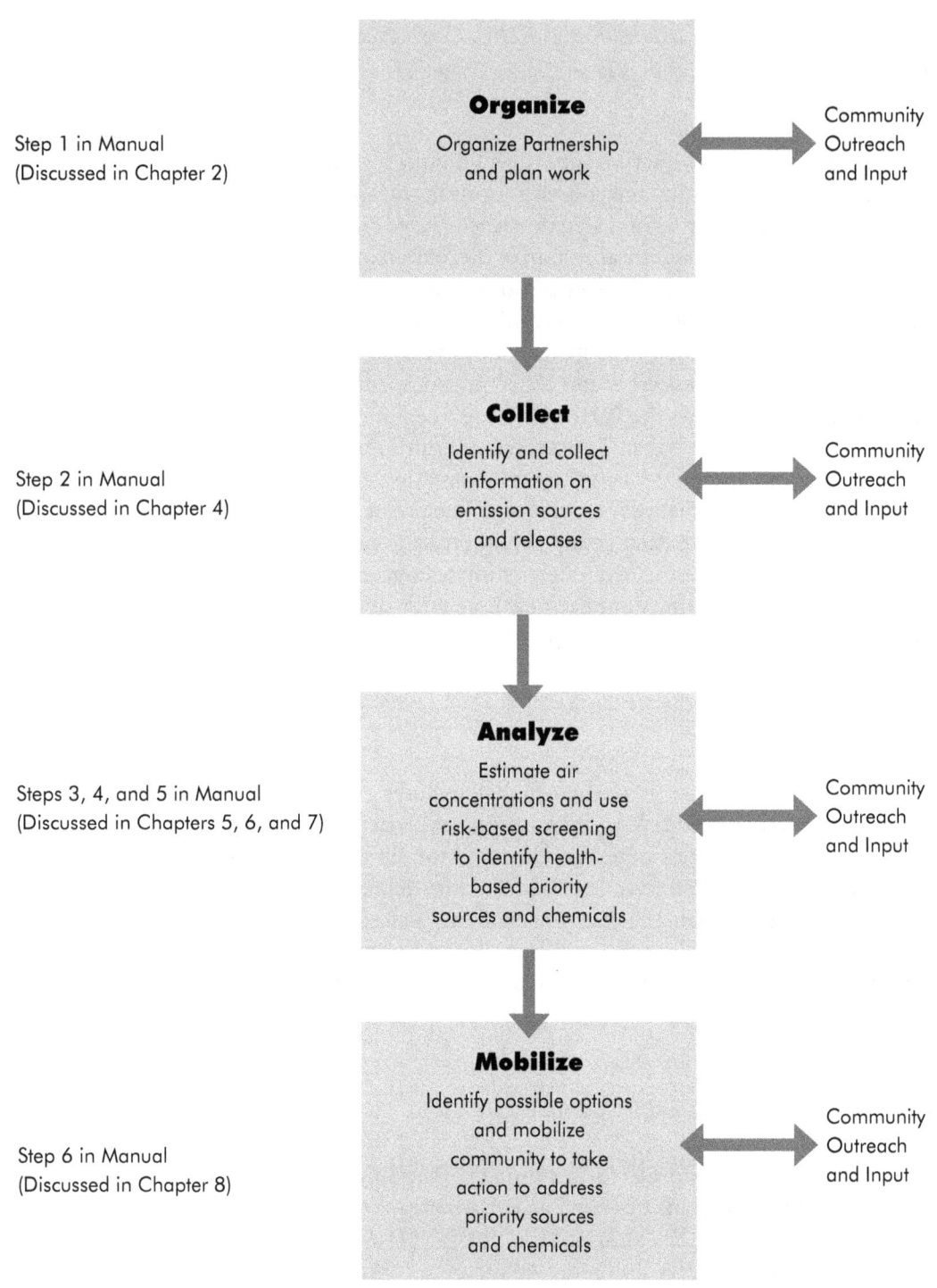

Step 1 in Manual
(Discussed in Chapter 2)

Organize
Organize Partnership
and plan work

Community
Outreach
and Input

Step 2 in Manual
(Discussed in Chapter 4)

Collect
Identify and collect
information on
emission sources
and releases

Community
Outreach
and Input

Steps 3, 4, and 5 in Manual
(Discussed in Chapters 5, 6, and 7)

Analyze
Estimate air
concentrations and use
risk-based screening
to identify health-
based priority
sources and chemicals

Community
Outreach
and Input

Step 6 in Manual
(Discussed in Chapter 8)

Mobilize
Identify possible options
and mobilize
community to take
action to address
priority sources
and chemicals

Community
Outreach
and Input

Figure 3-1.
Summary of Air Screening Methodology for the Community

be the source of the local knowledge, resources, and technical skills needed to complete all steps of the screening process and to implement the plan for improving local air quality. It will also serve as a community forum for participants to share ideas and to develop the trust and agreement that will be necessary to take effective action. The first step of the Partnership will be to clarify the goals of the project and develop a work plan to accomplish the community's goals. The Partnership will also need to develop a clear plan for communicating with the community at large, both to keep them informed on progress and to solicit their input as the work progresses. If the Partnership decides to use risk-based screening to analyze local air quality, then the following five steps will be used.

Step 2: Build an Inventory of Air Pollution Sources (Overview in Chapter 4; Technical Guidance in Chapter 9)

Once the Partnership is established and goals are set, the first task will be to identify and collect information on all the sources of air pollution in and around the target community. Local residents, businesses, and technical members of the Partnership will work together to identify and locate local sources such as local business and industrial facilities, power plants, landfills, cars, trucks and buses, household heating, burning of trash, municipal incinerators, etc. Information on the location, type, and amount of pollutants released from each of the sources will be collected or estimated. The Emission Source Inventory, based on the detailed information on local sources, will be unique for each community. Once completed, the source inventory will provide the community with a comprehensive database containing most of the information necessary to complete the screening process.

Step 3: Conduct the Initial Screen (Overview in Chapter 5; Technical Guidance in Chapter 10)

The Initial Screen begins the review of the information in the inventory to find if there are any releases that result in concentrations in community air that are above the community screening levels. To start this work, the Partnership will need to choose the screening level that will be used throughout the screening process. An evaluation of scientific information and community values and goals will form the basis for this choice. The Partnership must also set the standards and procedures that the technical members of the Partnership will use to select the information that will be used in the analysis. Setting these standards will ensure the quality of the technical work and legitimacy of the results.

To complete the first step in the screening process, a community will need to review a large number of chemicals and releases. Estimating the concentrations in community air that result from all these releases can be very resource intensive. To minimize the resources needed, the first screening step uses simple calculations and a look-up table to estimate the concentrations. The calculations and look-up table are designed to be conservative, which means that they are designed to overestimate the concentrations that would result from the releases. Conservative estimates of releases, meaning overestimates, are used at this step, and this also contributes to the overestimation of concentrations. Comparing these conservative overestimations to the community screening levels gives the community Partnership an easy way to set aside many chemicals and releases that result in concentrations equal to or below the community screening levels. (For the subsequent screening steps, the simple calculations and look-up table are replaced with more accurate methods of estimating concentrations. With fewer chemicals left to review in the Secondary and Final Screening steps, the Partnership will be able to use better, and more resource-intensive, methods to estimate concentrations.) When the Initial Screen is completed, the community will be left, it is hoped, with a shorter list of the chemicals and releases with concentrations above the community screening levels. These will be kept in the process for further analysis. It is important to remember and communicate clearly to the community that the list of chemicals that come out of the first screening should not be identified as chemicals of concern. Since the calculations and look-up table overestimated the concentrations, further and more accurate analysis may show that they are not a concern. These chemicals remain on the list for further analysis, but at the Initial Screening step it is still too early to identify the results as community priorities or concerns.

Step 4: Conduct the Secondary Screening (Overview in Chapter 6; Technical Guidance in Chapter 11)

The Secondary Screening starts with the chemicals identified in the first screening step. These chemicals and releases are analyzed using the same method, only this time, in place of the simple calculations and the look-up table used in the first step, air dispersion modeling is used to provide more accurate estimates of the concentrations in community air. These more accurate estimates of concentrations are then compared again to the community screening levels. Chemicals with concentrations equal to or below the screening levels are set aside; chemicals with concentrations above the community screening levels are kept for further analysis. For the Secondary Screening, the same conservative estimates (overestimates) of releases used in the first step are again used to complete the air dispersion modeling. As a result, the estimated concentrations in community air are, again, likely to be overestimates, so the chemicals will need further analysis to determine if they should be identified as community priorities.

Step 5: Conduct the Final Screen (Overview in Chapter 7; Technical Guidance in Chapter 12)

In the Final Screen, concentrations for the chemicals identified for further analysis in the second screen are once again compared to community screening values, only this time the most accurate information available is used to estimate the concentrations. With a limited number of chemicals left to review, the Partnership will have the resources to collect detailed information on releases from all of the sources of the chemicals reviewed in the Final Screen. This more accurate release information replaces the conservative estimates used in the first and second screens, resulting in the best possible estimates of concentrations in community air. Each of the sources can be contacted and asked to provide detailed information on their releases. State, tribal, and local agency files and databases can also be searched for more accurate information on the releases under review. Details on release amounts, location of releases, height of release stacks, etc., are collected and used with air dispersion modeling to provide the best possible estimates of concentrations. Concentrations estimated in this final step that are still above the community screening values are identified as priorities to be evaluated for possible community action.

Step 6: Develop Recommendations and Communicate Results (Overview in Chapter 8)

Once the priority chemicals are identified, the Partnership focuses on developing recommendations for potential actions that can be taken to reduce the impact of these chemicals on community health. Input from all the members of the Partnership will be needed to develop these recommendations. If they are not already members of the Partnership, any industrial, commercial, and institutional sources for the priority chemicals are invited to join the Partnership to help with

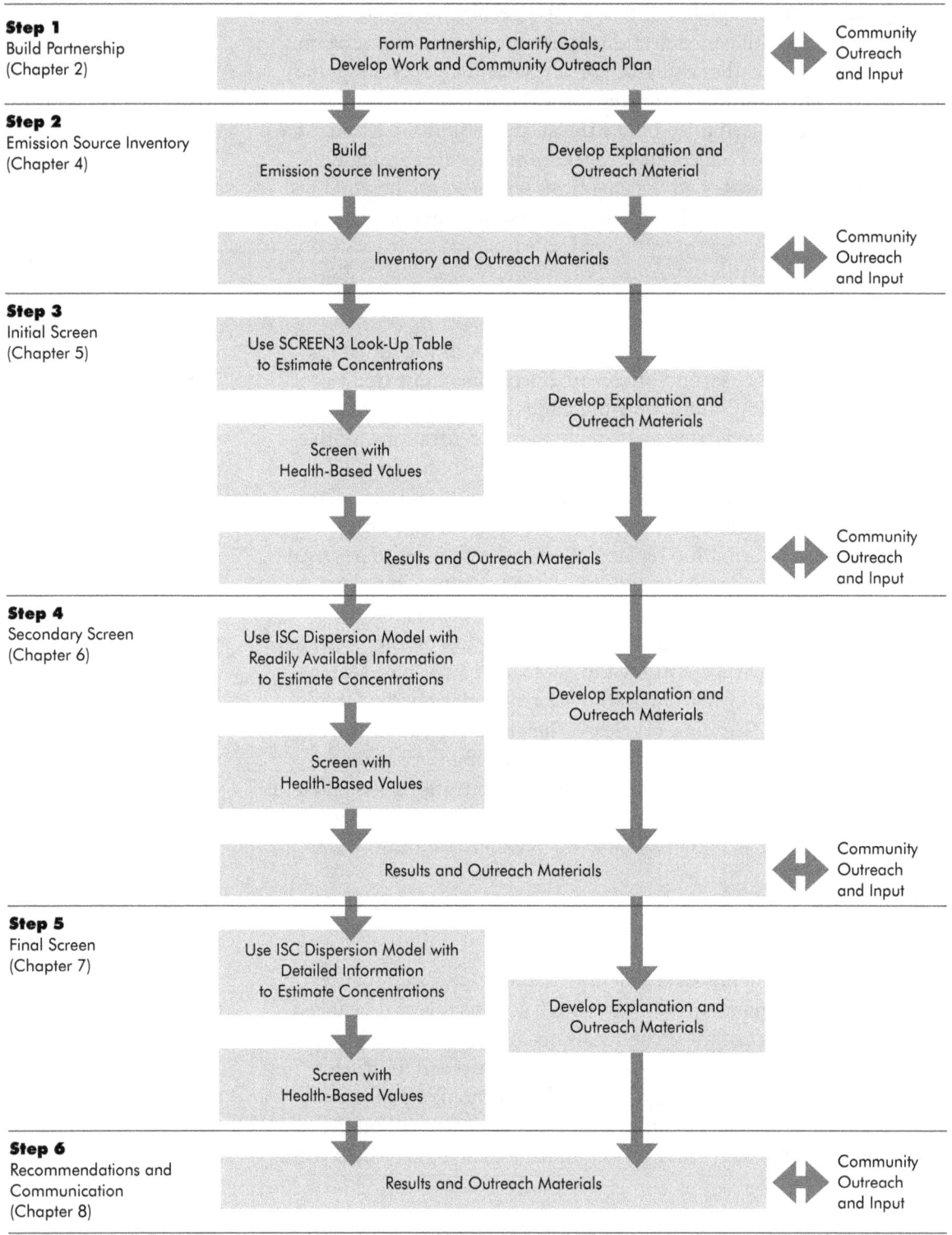

Step 1
Build Partnership
(Chapter 2)

Step 2
Emission Source Inventory
(Chapter 4)

Step 3
Initial Screen
(Chapter 5)

Step 4
Secondary Screen
(Chapter 6)

Step 5
Final Screen
(Chapter 7)

Step 6
Recommendations and
Communication
(Chapter 8)

Form Partnership, Clarify Goals,
Develop Work and Community Outreach Plan

Build
Emission Source Inventory

Develop Explanation and
Outreach Material

Inventory and Outreach Materials

Use SCREEN3 Look-Up Table
to Estimate Concentrations

Develop Explanation and
Outreach Materials

Screen with
Health-Based Values

Results and Outreach Materials

Use ISC Dispersion Model with
Readily Available Information
to Estimate Concentrations

Develop Explanation and
Outreach Materials

Screen with
Health-Based Values

Results and Outreach Materials

Use ISC Dispersion Model with
Detailed Information
to Estimate Concentrations

Develop Explanation and
Outreach Materials

Screen with
Health-Based Values

Results and Outreach Materials

Community
Outreach
and Input

Figure 3-2.
Detailed Community Air Screening Methodology

the development of recommendations. The Partnership also needs to communicate the results of the screening and its recommendations to the broader community. A final report, presentations to community groups, press releases, and other forms of communication can be used. This communication of the results and recommendations is the basis for mobilizing the community for the work that will be needed to carry out the recommendations to improve air quality. Figure 3-2 presents a flow chart summarizing all these steps and the relationship of this work to the community.

What resources would a community need to complete the process described in this Manual?

This Manual is designed to help a community partnership understand and improve air quality. All the work described in the Manual is designed to be done by the Partnership—not by someone or some institution working for the Partnership. The effort to build a partnership and trust, collect and analyze data, write and communicate results, and develop and implement plans for making improvements will require a significant commitment of time and resources to complete. Improved long-term community capacity to understand and address local air quality issues can justify this effort, but a community must recognize and plan adequately for this work to be able to complete the process and improve air quality. Please refer to Chapter 2 for a fuller discussion of the resources needed for building a partnership.

All the work described in this Manual is designed to be done by the Partnership—not by someone or some institution working for the Partnership

Past experience shows that it may take anywhere from one to two years, depending on the resources and size of the community, to complete the risk-based screening analysis and to develop and begin to implement recommendations for action. Remember that because of the time required to adequately understand local air quality, it will be important to find ways to take some actions to improve air quality as the screening process proceeds to reduce known risks, maintain community interest, and satisfy members of the Partnership more interested in action than analysis. As suggested in Chapter 2, using available information on typical community air risks may be sufficient for the Partnership to agree to take action on some likely high-risk sources and chemicals while simultaneously using the risk-based screening analysis to get a better understanding of the community's sources and priorities.

With the overall time frame in mind, the following is a checklist of resources that may be needed to carry out the process described in this Manual.

Resources Checklist

✔ An organization or person to champion and to facilitate the formation of the Partnership to start this effort. This could be a community or business leader or organization, a local school, or a government.

✔ A core committee of partners, with up to 20 members, willing to meet regularly for the length of the project and commit significant time to organize and implement the work. The core group should be a broad partnership committee with representation from as many community sectors, organizations, and governments as possible. The expertise needed for the project will come from this Partnership.

✔ A person devoted to organizing and sustaining community involvement. This person will support the volunteer community members of the core committee and help communicate the work of the core committee to the broader community. Community volunteers on the core committee will likely be participating in the Partnership after working on their regular full-time jobs, so they will need this support to effectively participate and reach out to the broader community.

✔ A trained facilitator. This facilitator, either from the community or brought in from outside, can concentrate on building trust and making sure the process works effectively.

✔ A regular meeting space.

✔ Access to a relatively new desktop computer to hold the community database.

✔ Access to an air dispersion model. Funding for this modeling may be necessary if the modeling capability is not available in the Partnership.

✔ Means to communicate effectively with the broad community: newsletter, flyers, meeting space for large meetings, printing and duplicating costs, etc. Existing means could be used or special ones created depending on community resources.

What skills will the members of the core Partnership group need to complete the process described in this Manual?

Please see the list of skills discussed in the Partnership section of Chapter 2. All the skills listed there will be needed to use risk-based screening to understand and improve local air quality.

What advance planning can be done to help ensure the success of the risk-based screening effort?

If possible, it would be good for the Partnership to get an understanding of the full process and do some advance planning and preparation so that it is prepared to complete each step of the effort as the work progresses. For example, the Partnership could identify all the teams that it will need to form over the course of the work and organize the teams early so that they can begin to assemble the resources and develop the skills they will need to carry out their tasks. The list of teams proposed in the Manual include:

• Communications Team	(Chapter 2)
• Emission Source Inventory Team	(Chapter 4)
• Quality Assurance/Quality Control Team	(Chapter 4)
• Concentrations Estimation Team	(Chapter 5)
• Screening-Level Concentration Team	(Chapter 5)
• Recommendations Team	(Chapter 8)

In the organization described in this Manual, each of these teams works for and reports to a core Partnership group or core committee with overall responsibility for the work. This form of organization helps to provide coordination and ensure that all the work meets community needs. But each Partnership will have its own needs and resources, so organizations will differ. This list of teams and their description in the Manual are offered only as suggestions to help begin the discussion of how to best organize the work in your community.

Please see the Partnership section of Chapter 2 for a discussion of other aspects of advance planning, including planning for fund-raising and for work with schools.

Do the community residents who are members of the Partnership need a background in science to participate?

No, the process is designed so that the science necessary to participate in the effort is provided to the residents and other non-technical participants as the project progresses. This Manual and the technical members who will provide the scientific expertise needed for the work of the Partnership will provide the science background that the non-technical members will need to participate fully in all decisions. In fact, all committee members will need to devote time to share information and help each other understand all the aspects of the community and the science that will be needed to reach the Partnership's goals of improving air quality.

Using the Manual will provide an important part of the picture of community risk, but not the complete picture

What are important limitations of this screening process that a community needs to understand before it starts?

While risk-based screening **does provide important benefits** to the community, it is important for the community to understand the limits of the methodology so that your expectations will be in line with the results you will get. A list of the limits, that is, things that cannot be accomplished with risk-based screening, is provided later in this section. As you read this list of things that cannot be accomplished, you may start to wonder why any community would want to adopt this approach. Providing this list of limits is not intended to discourage you. It is presented here to make sure that you understand up-front the things that using this Manual can and cannot accomplish. Some of the limits described later in this section flow directly from limits in current science or gaps in available information. Others are a result of the Manual's focus on a particular part of the complex set of factors that influence community health. Using the Manual will provide an important part of the picture of community risk, but not the complete picture. As discussed in the section on setting goals in Chapter 2, some partnerships may decide to supplement the information provided by using this Manual so that they can broaden the scope of concerns that they address. For example, for a better overall picture of risks from air toxics, the Manual recommends a consideration of other aspects of air quality, such as the acute effects of outdoor air toxics and the risks from indoor air exposures. Understanding clearly the overall picture that this method fits into will help your Partnership and community understand how to use and, if necessary, supplement the information this Manual provides. To help form a balanced picture, a summary of the results that can be obtained from the use of this Manual is provided following the list of limits.

A good way to begin to understand the limits of the method used in this Manual is to start with a discussion of the Manual's relationship to questions about community health and about illnesses that may have occurred in the community. The risk-based screening method used in this Manual cannot be used to establish the cause of current instances of disease in the community. Illnesses that have occurred in the community may have resulted from environmental exposures; or they may have been caused by non-environmental factors such as stress, diet, tobacco smoke, or genetics; or they may have resulted from a combination of environmental and non-environmental factors. This Manual focuses on current exposures to chemicals in outdoor air, so it considers only one among many potential causes of community illness. In addition, the method used in the Manual looks at **current** air quality, so it provides information on exposures that may result in illnesses that would occur **in the future.** Current illnesses are the result of exposures and other factors that occurred in the past. And, since the analysis looks only at outdoor inhalation exposures, it does not include all the environmental exposures that may affect community health, such as exposures to biological contaminants in the air, or exposures to pollutants in food and drinking water, or in indoor air and at the workplace, or to lead paint in the community's homes. So, as a result of its focus on outdoor inhalation exposures, the method in this Manual is not sufficient to identify causes of community illness. Using the Manual provides the information a community needs to identify and to use risk information to prioritize chemicals in outdoor air so that effective actions can be taken to reduce outdoor air risks. **To more fully address community health concerns, the Partnership would need to broaden its scope to consider more of the factors that may affect community health.** The work to understand outdoor air quality, the focus of the Manual, is only one piece of a broader community effort needed to understand and improve community health.

And, as discussed in the section describing work plan options in Chapter 2, the risk-based screening methodology does not include all the necessary analysis to assess the risk in the community that results from the chemicals in community air. For example, the methodology does not attempt to fully measure exposures in the community. A detailed exposure study would consider time spent in the neighborhood, activity patterns, population characteristics such the number of men and women and adults and children, etc. The information needed to set risk-based priorities does not require these measures, so to save resources, guidance for assessing risk is not included in this Manual. If a community is concerned about the levels of exposure and risk in the community due to outdoor air quality, a more detailed exposure assessment could follow this screening exercise. Suggestions for how to supplement the analysis described in this Manual to measure exposure and assess risk are provided in Chapter 7.

Because the methodology does not assess risk, it cannot provide information on the cumulative effects of multiple chemicals or all chemicals in community air. To get information on cumulative risk, the community would need to complete the exposure assessment for some or all of the chemicals in its inventory.

And, except in some limited cases, this screening analysis does not provide information on the possible effects of the mixtures of different chemicals in community air. It is possible that chemical mixtures could have more or less risk than the sum of the risk from each chemical in the mixture considered separately. The science to understand the effects of chemical mixtures is not very well developed.

In addition, all forms of risk analysis, including the risk-based screening used in the Manual, cannot, with certainty, account for all of the risks associated with the chemicals found in a community's air because some hazards may be unknown. Toxicity information for some chemicals may be unavailable or incomplete. New testing, such as the testing for effects on children and for effects on endocrine systems, may identify additional hazards that are currently unknown. Given the limits of the currently available toxicity information, the methodology is a review of *known* hazards, not all hazards. To take this into account, the Manual recommends a periodic review of community air quality so that any new information on the toxicity of chemicals can be used to adjust community priorities, if necessary.

And, since the methodology used in the Manual looks at only a part of the overall air quality picture, it may also be helpful to list the aspects of air quality that are not covered by this Manual. The methodology does not include indoor air and indoor air sources or short-term effects of chemicals in the outdoor air. Exposures resulting from air deposition, such as exposure that are the result of eating fish contaminated with mercury resulting from air releases, are also not considered. Nor does the method account for chemical transformations that may take place in the air after chemicals are released. Some chemicals break down rapidly and others are transformed into new toxic chemicals by reactions that take place in the atmosphere. None of these chemical transformations and their results can be accounted for with the method used in this Manual. In addition, aspects of air quality, such as visibility and dust, that are not related to toxic chemicals are not included in the analysis. All of these aspects of air quality mentioned above are important considerations for developing community priorities and community plans to address air quality. Information on each of these aspects of air quality can be found, and the Partnership will need to consider supplementing the information provided by using the Manual to develop a clearer overall picture of local air quality.

So, as a review, what part of the community environment **does** this Manual address and what can it do to help communities improve their health? In sum, this Manual provides guidance to help communities understand and address the chronic health effects that may result from breathing toxic chemicals in the community's outdoor air. The Manual helps communities identify the chemicals, and the sources of those chemicals, in outdoor community air that have the greatest potential to affect community health and that present a risk above a screening level set by the community. This information allows communities to set priorities and focus their efforts to improve outdoor air quality where they will do the most to reduce health impacts.

In more detail, a community can expect the following results from using this Manual:

- An inventory of all the significant sources of pollutants in your community's outdoor air

- Estimates of the combined concentrations of chemicals in community air resulting from all sources, and a comparison of these concentrations to risk-based screening levels set by the community

- Clear risk-based priorities for focusing community efforts where they can do the most good to improve outdoor air quality

- A partnership and better agreement in the community on priorities as the basis for mobilizing resources to make improvements

- A set of recommendations for improving local outdoor air quality

- Improved long-term capacity of the community to understand and improve local air quality

Building the Emission Source Inventory

What is the Emission Source Inventory?

The Emission Source Inventory is a collection of information on all the known sources of air emissions in and around the Partnership area. To start, the inventory will contain information on the amount of the releases of all the chemicals from all the sources that can be identified in and around the Partnership area and information on the location of all point sources. Eventually it will contain all the information on sources that the Partnership will need to complete all the steps of the air screening exercise described in this Manual. The inventory will serve as a permanent community database that can be updated periodically and used to measure progress in improving community air quality.

The Emission Source Inventory and all of the information that it will contain on local emission sources can also be used as a tool to educate the Partnership and the community. This educational work on the local sources of air pollution will lay the foundation for the future work of the Partnership to improve air quality.

This chapter gives an overview of the work needed to set up the database, identify and locate stationary point sources, and collect the information on the amount of the releases from all sources. Additional instructions for collecting the information needed for each screening step can be found in the subsequent chapters of the Overview and in the Technical Guidance sections of the Manual.

Organize
Collect
Analyze
Mobilize

How is the Emission Source Inventory used in the risk screening exercise?

As described in the Introduction, the screening methodology used in this Manual contains guidance for developing estimates of concentrations of chemicals in community air and guidance for comparing these concentrations to community screening-level concentrations to identify priority chemicals. The information on sources and air emissions collected in the inventory provides the basis for the Partnership's work to develop the estimates of concentrations of chemicals in community air. Once completed, the Emission Source Inventory will contain all of the source information needed to estimate concentrations for all the steps of the screening process.

How does the Partnership get started in developing the Emission Source Inventory?

The first step is to define your study area, that is, the area that will be the focus of the Partnership's air screening effort. The boundaries of the study area should, if possible, follow existing boundaries, such as the boundaries defining neighborhoods or community organizations. Adjusting these existing boundaries so that they fit the way data are organized in state and federal air quality databases (e.g., ZIP codes) may be helpful in collecting the information. In Baltimore, the study area was defined by five adjacent Baltimore neighborhoods that were located in two primary ZIP codes, 21225 and 21226. The residents in those neighborhoods joined together with local, state, and federal government agencies to form the Community Environmental Partnership.

Is there a need to include sources beyond the study area boundaries?

Sources outside the study area can have an impact on the air quality within the study area depending on their strength (emission quantity) and the meteorological conditions, such as the prevalent wind speed and direction in your area. For example, if your study area is downwind from a large source like a power plant or refinery, then there is the potential for air quality to be affected by this source. For this reason, it should be considered for inclusion in the analysis. Similarly, a downwind source may not affect the study area and may not need to be included in the analysis. Both large and small sources that are very close to the Partnership area will probably need to be included.

To decide on how far out from the study area to go and which facilities to include, the Partnership will need to rely on the professional judgment of its technical members. This decision will be based on review of all the factors needed to determine the potential of a source to affect the Partnership's study area. For example, in Baltimore a decision was made to include all large facilities within five miles of the Partnership area. This added eight additional ZIP codes (i.e., 21060, 20161, 21090, 21122, 21219, 21222, 21227, 21230) to the primary ZIP codes (i.e., 21225 and 21226).

The Partnership may also want to consider adding especially large emitters on an individual basis that are outside the boundaries the Partnership sets for including large facilities. Information on these large sources, such as a large power plant or industrial facility, could come from members of the committee or from a search of databases such as the National Emissions Inventory (NEI) or the Toxics Release Inventory (TRI). In Baltimore, this was considered, but no such facility was identified.

Will all emissions be captured by this study?

While the goal is to capture as many of the sources as possible, it may not be possible to account for all the emissions that affect the Partnership area. Examples of situations in which emissions and chemicals might be missed are provided below. Partnerships will need to consider their resources and goals to decide how much effort to expend on identifying all possible sources.

Missing sources: Some sources, such as small businesses operating without permits, may not be included in the release and permit databases maintained by federal, state, tribal, and local governments. The Partnership will make use of all of its members to organize an effort to identify these kinds of sources, as discussed later in this chapter, but some small sources may still be missed.

Missing chemicals: Even if all the sources are accounted for, one or more of the sources may release a chemical that is not included in any of the existing databases and, as a result, the chemical and its releases could be missed. The Partnership will need to rely on the knowledge and expertise of its members to try to identify all of the chemicals that need to be included in the Partnership's Emission Source Inventory. It may be helpful, if sufficient resources are available, to check the list of chemicals stored in local facilities, which is maintained by the Local Emergency Planning Committee, to see if there are any chemicals that should be added to the Partnership's inventory. And facilities, especially large facilities likely to use or release multiple chemicals, can be asked to report on any releases they might have of chemicals not included in the existing databases.

Missing releases: Even if all sources and chemicals are accounted for, the Partnership may also miss some releases of these chemicals if it relies only on the information in the existing release and permitting databases. Accidental releases and releases due to maintenance may occur and are sometimes reported in special databases or files kept at the local or state level. It is also possible that some accidental, maintenance, and even regular production releases may not be reported at all. Input from committee members with a detailed knowledge of the study area, along with state, local, and tribal government staff with an understanding of all the information collected and stored at the state and local level, will be crucial for including as many significant releases as possible in the inventory. Please keep in mind that short-term releases, such as accidental and maintenance releases, are not likely to change the annual average release amounts used to identify the priorities in the screening method used in this Manual. Periodic peak releases, either accidental, maintenance, or production, are more likely to be of concern for understanding short-term acute effects. Although these acute effects are important to the community, it was not possible to include an analysis of them in this Manual.

Secondary formation of chemicals: In some cases air pollutants are formed by chemical reactions that occur in the atmosphere with some of the chemicals that are released. Studies have shown that the secondary formation of chemicals (e.g., formaldehyde, acetaldehyde, and acrolein) in the atmosphere, especially from chemicals released by mobile sources, is an important source of chemicals in urban air. The methods used to estimate this secondary formation of chemicals in the atmosphere are complicated and beyond the scope of the screening methods included in this Manual. The Partnership can include chemicals formed in the atmosphere in the screening process if local monitoring data for these chemicals are available. Information on these chemicals and their concentrations can also be found in the National Air Toxics Assessment (NATA) database. The NATA database is available at http://www.epa.gov//ttnatw01/nata/.

What skills and resources are needed to complete the Emission Source Inventory?

The Partnership will need skills and resources in the following areas to complete its inventory and communicate its work to the community:

- Familiarity with the data available at the local, state, tribal, and federal levels on sources and emissions

- Knowledge of the community and the sources of emissions in the community

- Community leaders to mobilize the community to assist with information collection

- Communication skills to inform the community on the sources of local air pollution

- Familiarity with database management or spreadsheet programs

- Ability to use estimation factors or other estimation methods to estimate emissions from sources when no data are available

- A desktop computer with Internet access and spreadsheet software

- Geographic Information System (GIS) skills to geographically illustrate and locate local sources of air pollution, if available

Staff from the state, tribal, or local departments of the environment will be familiar with the state, tribal, and national databases that contain information on sources. Those agencies may have already assembled a comprehensive air emission inventory for the National Toxics Inventory. This will be an excellent starting point for constructing the Emission Source Inventory for the study area.

Members of the Partnership from governments, environmental organizations, industry, or academic institutions and community professionals will have the technical skills needed for accessing and using databases and for estimating emissions.

Resident and industry members of the Partnership will have the detailed knowledge of the study area to help finalize the inventory through "ground-truthing," identifying sources that may have been missed as well as ones that have gone out of business. And community leaders in the Partnership will be able to help mobilize the community to collect any information not available in the databases.

The Emission Source Inventory can be stored on a computer using spreadsheet software. In the Baltimore project, the inventory was stored using a commercially available spreadsheet. The Emission Source Inventory computer should have access to the Internet so that publicly available databases can be accessed.

The skills needed for communicating the inventory work to the broader community can come from the different members of the Partnership working on the communication team.

How will the Partnership collect the data?

Only the knowledge and a broad level of participation of local residents and businesses will ensure that as many sources as possible are identified and included in the inventory

The Partnership will probably want to appoint a technical team from among its members to plan and carry out the work of completing the Emission Source Inventory. Ideally, this team, referred to as the Inventory Team in this Manual, would include both technical members of the Partnership and non-technical community members interested in collecting information on local sources. The Inventory Team will be needed to collect new information at each step of the screening process, so its work will continue throughout the project. The work of the Inventory Team will range from assembling information from databases for the Initial Screen to organizing members of the Partnership and community to visit sites to collect the detailed information that will be needed in the Final Screen.

Please remember that although the Inventory Team will be able to collect much of the data on its own, it will need to rely on the resources of the full Partnership and the community to assist in the data collection and to review the information collected to ensure that it is accurate and complete. For example, only the knowledge and a broad level of participation of local residents and businesses will ensure that as many sources as possible are identified and included. The full Partnership committee will also have to explain the Emission Source Inventory to the broader community. Therefore, it will be important for the entire Partnership committee to understand the work of its Inventory Team and the variety of information that the Partnership will collect and store in the Emission Source Inventory database. The information that all the members of the Partnership will need to understand and to help build the Partnership's Emission Source Inventory are provided in this and subsequent chapters of the Overview. Detailed technical guidance to help the Inventory Team collect the information needed to complete the screening is provided in the Technical Guidance section of the Manual.

What will the Inventory Team need to do to plan for the work to complete the Emission Source Inventory?

In addition to setting up the inventory database and collecting information on releases and stationary point source locations, the Inventory Team will also need to develop a plan for collecting the rest of the information that will be needed to complete all of the screening steps. To develop this plan, the Inventory Team will need to review all of the chapters of the Manual to get an understanding of the information that will be needed. Table 4-1 summarizes the source types and information that needs to be collected to complete the risk-based screening (i.e., Initial, Secondary, and Final Screening steps). Explanations of the information listed in this table will be provided in this and subsequent chapters of the Overview. The Inventory Team will also need to work closely with the Partnership team that has the responsibility for using the information in the inventory to estimate concentrations. These two Partnership teams will need to make sure that they share the same understanding of the information that will be needed. In addition, some of the information requirements will depend on the decisions that the team responsible for estimating concentrations will make. For example, for the Initial Screen, the team estimating concentrations will have a choice of methods to use, and the information the Inventory Team will be required to collect will depend on this choice. Details of this choice will be explained in the Initial Screen Overview and Technical Guidance section.

The Inventory Team's plan for collecting information will also have to take into account the fact that the team will not know in advance which chemicals will be identified at each step of the screening process. For example, detailed information on stack releases, such as the temperature of the gases on release, will be needed to estimate concentrations in the Final Screen. But collecting this information for all the stack releases in the inventory would be a waste of time, since the information will only be needed for the small number of chemicals left after the Secondary Screen is completed.

Despite the fact that the Inventory Team will need to wait for the screening to proceed to know which chemicals to collect information for, it is still important for the team to understand all the data that will be needed to complete the screening. Understanding all the data needs will allow the team to take advantage of easy opportunities to collect information if they arise. For example, if all the detailed information on releases that will be needed in the Final Screen comes in the same database used to collect the information on the amount

Table 4-1.
Summary of Information Needed to Complete Risk-Based Screening

TYPE OF SOURCE	INITIAL SCREEN INFORMATION NEEDED	SECONDARY SCREEN INFORMATION NEEDED	FINAL SCREEN INFORMATION NEEDED	SOURCE OF INFORMATION
STATIONARY SOURCES: **Stationary Point Sources:** All large and small commercial, industrial, and public facility sources (sources will be modeled individually)				
• Sources with release information available	• Chemicals released • Release amounts • Distance to nearest receptor • Release type: stack or fugitive • Approximate stack height	• Chemicals released • Release amounts • Source location • Modeling release parameters • (Local meteorological data)	• Detailed release information provided by facilities	• State, tribal, local, and federal databases • Local knowledge • Business databases • Facility contacts for detailed business or release information
• Sources with release information *not* available	• Chemicals released • Release amounts • Distance to nearest receptor • Release type: stack or fugitive • Approximate stack height • (Type of business) • (Chemical use or business activity measure) • (Emission factors)	• Chemicals released • Release amounts • Source location • Modeling release parameters • (Local meteorological data)	• Detailed release information provided by facilities	• State, tribal, local, and federal databases • Local knowledge • Business databases • Facility contacts for detailed business or release information
Stationary Area Sources: Household, small office building, and other miscellaneous sources (sources combined for modeling)	• Total release amounts for county	• Chemicals released • Total county emissions • (Local meteorological data) • (Census tract geographical information and centroids) • (Information needed to apportion county emissions)	• Same as Secondary Screen	• National Emissions Inventory (NEI) and NATA databases • State databases • Business databases • Local knowledge

Note: Information required for screening, but not included in the Emission Source Inventory, is listed in parenthesis.

of releases, and if the information is in a form that will allow it to be transferred electronically into the Partnership's inventory, then detailed information on all the chemicals can be collected from the beginning with minimal effort. Even if the detailed information is not

Table 4-1.
Summary of Information Needed to Complete Risk-Based Screening (continued)

TYPE OF SOURCE	INITIAL SCREEN INFORMATION NEEDED	SECONDARY SCREEN INFORMATION NEEDED	FINAL SCREEN INFORMATION NEEDED	SOURCE OF INFORMATION
MOBILE SOURCES: On-road	• Not estimated in Initial Screen (information needed for Secondary Screen collected for inventory)	• Chemicals released • Total county emissions • (Local meteorological data) • (Census tract geographic information and centroids) • (Information needed to apportion county emissions)	• Traffic counts • If necessary, information needed for Mobile6 mobile source emission model and ISCST line source modeling	• Local knowledge • Transportation planning organization
Non-road	• Not estimated in Initial Screen (information needed for Secondary Screen collected for inventory)	• Chemicals released • Total county emissions • (Local meteorological data) • (Census tract geographic information and centroids) • (Information needed to apportion county emissions)	• Detailed information for large, non-road sources, if any	• Local knowledge • State, tribal, and federal modeling expertise
BACKGROUND SOURCES:	• National background concentration measurements	• Same as Initial Screen	• Same as Initial Screen	• State, tribal, and local monitoring authorities • Monitoring databases
MONITORED CONCENTRATIONS:	• Annual average concentrations	• Annual average concentrations	• Annual average concentrations	• Local, state, tribal, and national monitoring databases

Note: Information required for screening, but not included in the Emission Source Inventory, is listed in parenthesis.

easy to collect, understanding that it may be needed will, at a minimum, allow the Inventory Team to note where the information can be found if it is needed at a later date.

As it carries out its plan to collect information, the Inventory Team will also need to report back to the full Partnership committee on any sources for which data are not available and let the Partnership know where the team will need help to collect missing information. The full Partnership committee can then help to mobilize the community to collect this information. For example, if the information about the number of vehicles using a road through the community is not available, the Partnership may ask a participating high school to organize a traffic survey.

How will the Partnership ensure that the work of its Inventory Team and all of the work of the Partnership is done accurately and according to the standards set by the Partnership?

Over the course of the work, the Partnership and its teams will collect and transfer detailed information on hundreds of releases, carry out hundreds of calculations to develop screening concentrations for all the chemicals, and set up and run computer models containing large amounts of information. Since the Partnership and the community will rely on all of this work to make decisions and to commit resources, it will be essential to ensure that all the work is done accurately and according to the Partnership's guidelines. Even small errors in transferring information from one database to another, such as the misplacement of a decimal point, could change the outcome of the screening process for a chemical. Since the teams may make some unintentional errors during the course of their work, some system for checking work and identifying errors will need to be established. Quality control methods for ensuring data quality and accuracy are available and the Partnership should consider the available resources and establish its own system to ensure the accuracy of the screening work. If enough resources are available, the Partnership may want to set up a separate quality control committee and give it the responsibility for reviewing the work of all the technical committees. Setting up a separate committee for data quality would allow the reviewers to be different from the persons doing the work. Teams checking their own work usually have a more difficult time spotting errors. If a separate committee is not possible, each team will have to set up a system to double check all of its work to make sure that it is accurate.

What are the different types of emission sources that may exist in the Partnership area?

There are a wide variety of air emission sources that need to be included in the Partnership inventory, ranging from large chemical facilities and power plants to the gasoline-powered lawnmowers that community members may use to cut their grass. The following is a list of the different kinds of sources that may be found in communities, with examples to illustrate each type of source. As is customary, this list is divided into stationary, mobile, and background sources.

Stationary sources: Stationary sources include all the emission sources that come from fixed locations. This type of source includes both large and small stationary sources, ranging from large sources such as electric utility plants, chemical plants, steel mills, oil refineries, and hazardous waste incinerators, to small stationary sources such as the neighborhood dry-cleaners and auto repair and refinishing shops. Small stationary sources also include all the emission sources from homes and small office buildings, such as wood stoves, gas-fired home heating, and even household chemical uses.

In this Manual, these stationary sources are divided into two separate groups based on the methods that will be used to estimate the concentrations that result from their releases. The first group of stationary sources includes all the sources whose concentrations will be estimated individually. All the large and small businesses, from large chemical facilities and steel mills, to the gasoline station on the neighborhood corner, are included in this group. Publicly owned facilities such as incinerators and water treatment plants are also put into this group. Each of these facilities will have the concentrations resulting from its releases estimated individually. For consistency and to avoid repeating the long description of this group, in this How-To Manual, all of these stationary sources (i.e., those whose concentrations will be estimated individually) are called stationary point sources.

The second group of stationary sources includes all the sources that are too numerous or irregular to handle practically as individual sources. Household sources such as heating furnaces, small offices, and exterior painting are examples of these sources. Because the releases from these sources are too numerous to have their concentrations estimated individually, they are combined together by type. For example, the releases from all the home furnaces are grouped together, and the concentration that results from the combined releases is estimated. In this Manual, all of these kinds of sources, that is, sources whose releases will be combined for estimation, are called *stationary area sources*. The sources that are treated as *stationary area sources* are listed here, using the categories established in the National Toxics Inventory:

- Agricultural production

- Asphalt paving: cutback asphalt

- Consumer products usage

- Gasoline distribution stage 1
- Industrial boilers: distillate oil
- Institutional/commercial heating (all types)
- Natural gas transmissions and storage
- Non-industrial asphalt roofing
- Pesticide application
- Residential heating (all types)
- Residential heat (wood)
- Structure fires
- Surface coatings: architectural
- Surface coatings: industrial maintenance
- Surface coatings: traffic markings

Mobile sources: Mobile sources include all the emission sources that move or can be moved from place to place. This group contains all types of vehicles and equipment, from cars and trucks to construction equipment and lawnmowers.

Databases with information on mobile sources are divided into two categories: *mobile on-road* and *mobile non-road.* The Partnership's inventory database will also use these categories to store its information on mobile sources. On-road mobile sources include all cars, buses, trucks, and any other vehicle used on streets and highways. The non-road mobile sources include everything from airplanes and ships to construction equipment and lawnmowers.

Background sources: There are three types of sources contributing to community air concentrations that are called background sources: very distant sources, such as the major power plants in the Midwest that contribute to air concentrations in Eastern cities; natural sources, such the production of methyl chloride by naturally occurring bacteria in bodies of water; and chemicals persisting in the environment from past human uses, such as the releases from soil contaminated with carbon tetrachloride, a chemical widely used in the past.

What information on these emission sources will the Inventory Team collect to start the Emission Source Inventory database?

As the Partnership prepares to conduct its screening the Inventory Team will focus first on collecting information on the amounts of releases from the emission sources and on identifying and locating stationary point sources. The additional information that will be needed to complete each step of the screening process is discussed in the subsequent chapters of the Manual. The release amount information that will be collected and entered into the inventory database for each kind of source is described below:

- Stationary point sources: the release amounts for each chemical from each source.

- Stationary area sources: the total emission amount for the county in which the Partnership is located for each type of area source.

- Mobile on-road sources: the total emission amount for the county for all on-road mobile sources.

- Mobile non-road sources: the total emission amount for the county for all non-road mobile sources.

- Background sources: Estimated background concentrations provided by the EPA Air Office. Universities may also have research information on local background concentrations. Information on background source concentrations may also be obtained by reviewing local monitoring data.

Table 4-2 summarizes the source types and the information that needs to be collected to start the Emission Source Inventory database. Information on the amount of releases for all of the large and some of the small stationary point sources will be available for each individual source in state, tribal, and national databases. For some small stationary point sources, the Inventory Team will be able to find only estimates of the combined releases for all of the sources in a category. For example, in some states, releases for dry-cleaning facilities are not available for each facility. Only an estimated total release amount for all the dry-cleaners in a county will be available. In that case, since all of the concentrations for all the stationary point sources need to be estimated individually, the Inventory Team will need to estimate releases for each of the dry-cleaners in the Partnership area. To estimate the releases from these sources, the Partnership may need to collect information such as the chemical use or sales volume of the business. Methods to estimate the releases based on these kinds of information are available, and directions on their use are provided for the Inventory Team in the Technical Guidance section of the Manual.

During the Initial and Secondary Screening steps, the Partnership will be using readily available release data based on maximum permitted release amounts as inputs for the look-up table and the ISCST model, respectively. If the release data available to the Partnership are only

Table 4-2.
Source Types and Information Needed to Start Building Emission Source Inventory Database

TYPES OF SOURCES	INFORMATION NEEDED TO START BUILDING THE EMISSION SOURCE INVENTORY DATABASE
STATIONARY SOURCES:	
Stationary Point Sources: All large and small commercial, industrial, and public facility sources (sources will be modeled individually)	
• Sources with individual release information available	• Release amounts and location for each source
• Sources with release information *not* available	• Location information plus additional information needed to estimate release amounts for these sources
Stationary Area Sources: Household, small office building, and other miscellaneous sources (sources will be combined for modeling)	• Total county release amount for each type of source
MOBILE SOURCES:	
On-road: All truck, buses, cars, and any other street or highway vehicle	• Total county release amount for all on-road sources
Non-road: Trains, airplanes, ships, construction equipment, lawn equipment	• Total county release amount for all non-road sources
BACKGROUND SOURCES:	
Releases from very distant sources and releases that are not the result of current human activity, including both natural and past human sources	• Concentrations for each background chemical

estimated actual release amounts and not maximum permitted release amounts for stationary sources, the Partnership will need to increase these release estimates, possibly multiplying the release amount by a factor agreed on by the Partnership, to ensure that the Initial and Secondary Screens are conservative. For example, release estimates taken from the TRI database are estimated actual releases, not maximum permitted releases, so the Partnership's technical team will need to increase these amounts to ensure that the Initial and Secondary Screens are conservative. For the Final Screening step the Partnership will have the resources to contact each of the remaining stationary sources to obtain actual release data.

What chemicals will be included in the Emission Source Inventory?

The goal of the Partnership will be to include as many chemicals as possible in its inventory. To do this, the Partnership will have to access several databases or information sources maintained by different government agencies. Each of these databases and data sources contains information on its own list of chemicals. Many of the chemicals will be the same in all of the data sources, but some chemicals will be found in only one and not the other databases. The Partnership will use all of the available data sources to ensure that it has information on as many chemicals released in its area as possible. Further guidance on accessing these data sources can be found below and in the Technical Guidance section of the Manual.

Since some of the releases from a source can be found in more than one database, the Partnership will have to be careful to avoid double counting releases. If the same releases are recorded in more than one database, the Partnership's Inventory Team will use only the information from the database with the most up-to-date information. It is also possible that some releases may have been entered more than once in the same database by error. If there are multiple releases of the same chemical from the same facility, the Inventory Team may want to check to be sure that they are actually different releases and not multiple entries for the same release. Checking to make sure that the release amounts entered into the Partnership's database are not the result of counting the same release more than once will be a key part of the Partnership's quality control work.

If monitoring information is available for the Partnership area, should it also be included in the Emission Source Inventory?

First, some brief background: Monitoring directly measures what is in the air either at fixed locations in a community or as the air is breathed by someone living in the community through personal monitors worn by community volunteers. Since monitoring relies neither on estimates based on computer modeling nor on release data, it can provide accurate measures of concentrations at specific locations. Setting up new monitoring is an appropriate and thorough way to analyze community air, but it can be expensive, and it will require at least a year of measurements for estimates of long-term effects. Because of its cost, monitoring can only be done in a limited number of locations in a community. The information on concentrations from monitoring also has some limits. By itself, monitoring does not provide information on the sources of the chemicals measured at the monitoring location. And most monitoring involves sampling air at regular intervals, so monitored concentrations can only represent the concentrations in the air at the times of measurement. The screening methodology described in this Manual uses existing monitoring information, if it is available, but, to minimize resources, it does not require new monitoring. The possibility of monitoring to verify the results of the screening exercise is discussed in Chapter 7.

In place of new monitoring, the screening method in this Manual makes use of the information from monitoring stations that have been set up and are operating in cities across the nation. If a monitoring station is located in or near the Partnership neighborhood, and the Partnership has determined that the monitoring data are appropriate for use, the Inventory Team should collect all the available information on monitored concentrations and enter the information into the Emission Source Inventory database. (Remember that monitoring stations measure concentrations of chemicals. This concentration information is different from the release information collected for the Initial Screen for each source, so it will need to be separated from the release data and labeled appropriately in the Emission Source Inventory database.) These concentrations will be screened along with the concentrations estimated with air dispersion models. The availability of appropriate monitored concentrations will allow the Partnership to verify the accuracy of its modeling. The monitoring can also be used to help ensure that all the sources of monitored chemical emissions have been identified. Guidance for using the monitored concentrations will be provided in subsequent chapters.

How will the Partnership decide if the available monitoring information is appropriate for use in its screening exercise?

This Manual encourages the use of ambient air monitoring data to supplement and verify concentrations estimated by air dispersion modeling. However, when considering how to use monitoring data, the team must understand the goals of the monitoring program collecting the data and the limitations in the spatial and temporal coverage of the data. For example, a program designed to determine the average concentrations of air pollutants across a city might not be an appropriate surrogate for the concentrations of pollutants at the fencelines of facilities releasing the chemicals.

Several things should be considered when deciding whether or not to use monitoring data instead of estimated airborne pollutant concentration.

- Does a monitoring network exist for the pollutants of interest and what are the averaging times of concern?

- What is the location of the air monitor in relation to the study area?

- Was any air modeling already done to locate the monitoring station?

- Will the local topography greatly influence the local air patterns and dispersion?

- Has the monitoring network been designed to locate points of maximum concentrations, average concentrations, or other?

- Do the data set and analysis allow the impact of the most important individual sources to be identified if more than one source or emission point is involved?

- Is at least one full year of valid ambient data available ?

These questions are intended to help in making the decision whether to use monitoring data on concentrations to supplement the concentrations estimated by air dispersion modeling. It is recommended that this discussion take place and that the decision on the usefulness of available monitoring data be made prior to entering the data into the Emission Source Inventory database.

What are sources for the information that will be needed?

To collect the information needed to complete the Emission Source Inventory database, the Partnership technical team will have to draw on a variety of different sources. In some cases, where information has not been collected or is not available, the Partnership may need to mobilize its members and the community to collect the information that will be needed. The following is a list of information sources for the data and information that will be needed to complete the inventory.

Local knowledge: Members of the Partnership and community, including residents and local businesses, will have a lot of information on the location and releases from sources in the Partnership area. This local knowledge will be used to "ground-truth" and supplement the information collected from the government databases discussed below. For example, local residents and businesses may be able to identify sources not listed in databases, releases not reported, or heavily used roads and unusual truck traffic. And in the cases where important information is not available, the Partnership can also organize the collection of information needed to complete the screening process such as traffic counts or information on small businesses.

State, tribal, and local government air permitting databases: The state, tribal, or local government authority responsible for permitting releases to air maintains a database with information on local sources and their releases. This is the primary source for government information on local air sources. The amount and kind of information collected will depend on the local authority. The state, tribal, and local databases, since they often contain the most up-to-date information available, are recommended as the preferred source of data for this screening exercise. If access to the database covering your state is not available, information can be obtained from the National Emissions Inventory, described next.

National Emissions Inventory (NEI): This database of air emissions information is prepared by EPA with input from numerous state and local air agencies, from tribes, and from industry. This database contains information on stationary and mobile sources that emit criteria air pollutants and their precursors, as well as hazardous air pollutants (HAPs). A list of the HAPs is shown in Appendix E. Criteria pollutants are used by EPA as indicators of air quality. The criteria pollutants are ozone (O_3), carbon monoxide (CO), nitrogen oxides (NOx), sulfur dioxide (SOx), particulate matter with diameter less than or equal to 10 micrometers (PM_{10}), particulate matter with diameter less than 2.5 micrometers ($PM_{2.5}$), lead (Pb), volatile organic compounds (VOCs), and ammonia (NH3). HAPs are generally defined as those pollutants that are known or suspected to cause serious health problems. Section 112(b) of the Clean Air Act currently identifies a list of 188 pollutants as HAPs, including solvents such as benzene, toluene, and xylene. More information about the NEI database and the compilation of criteria pollutant and HAP emissions inventories, and links to the database, are available on the CHIEF NEI web page at http://www.epa.gov/ttn/chief/net.

The list of HAPs is available at http://www.epa.gov/air/urbanair/6poll.html. The list of HAPs is also provided in Appendix E.

Toxics Release Inventory (TRI): The Toxics Release Inventory (TRI) is a publicly available EPA database that contains information on toxic chemical releases and other waste management activities reported annually by certain covered industry groups as well as federal facilities. This inventory was established under the Emergency Planning and Community Right-to-Know Act of 1986 (EPCRA) and expanded by the Pollution Prevention Act of 1990.

The TRI program has expanded significantly since its inception in 1987. The Agency has issued rules to roughly double the number of chemicals included in the TRI to approximately 650. Seven new industry sectors have been added to expand coverage significantly beyond the original covered industries, that is, manufacturing industries. Most recently, the Agency has reduced the reporting thresholds for certain persistent, bioaccumulative, and toxic (PBT) chemicals in order to be able to provide additional information to the public on these chemicals. Information on the TRI is available at http://www.epa.gov/tri/.

Ambient air monitoring databases: Federal, state, tribal, and local agencies sometimes establish monitoring stations to directly measure the levels of chemicals in the air. This monitoring will be increased over the next several years as a result of the new National Air Monitoring Program established by the Clean Air Act. Information on the measurements of chemicals at monitoring stations in or near the Partnership area may be available. Government staff members of the Partnership will know how to access this information. For information see http://www.epa.gov/ttn/amtic/stratmem.html.

Local transportation planning organizations: The information needed to estimate concentrations for mobile sources may be available from the organizations that plan for local transportation. As a part of their efforts to plan for transportation needs, these organizations are required to estimate mobile source emissions for urban metropolitan areas. Roadway information and the number of vehicles traveling on the roads may be available from this source.

National Air Toxics Assessment (NATA): The NATA program provides background concentrations based on monitored values identified in the Cumulative Exposure Project, a study that estimated nationwide 1990 ambient concentrations of air toxics. Based on that study, nationwide background concentration values for 13 toxic air pollutants were developed. Details for including these background concentrations in the Emission Source Inventory can be found in Chapter 9, the Technical Guidance chapter for building the Emission Source Inventory. Additional information on these background concentrations can be found at http://www.epa.gov/ttnatw01/nata/backcon.html.

Commercial business databases: There are a number of commercial databases that list and describe business activity for every part of the nation. These sources of information can be used to help find and identify all the businesses in the area that may be using and releasing chemicals. These data sources may also contain information on the size and activity of the businesses. If release information for small businesses is not available, this information may help the Partnership technical team to develop release estimates for these businesses. In Baltimore, the Partnership used the commercially available Dunn and Bradstreet database to help identify area businesses.

Chemical Emergency Preparedness and Prevention Office (CEPPO): EPA's Chemical Emergency Preparedness and Prevention Office provides leadership, builds partnerships, and offers technical assistance to prevent and prepare for chemical emergencies, respond to environmental crises, inform the public about chemical hazards in their community, and share lessons learned about chemical accidents. Their web site is at http://www.yosemite.epa.gov/oswer/ceppoweb.nsf/content/mission.htm.

They maintain a Local Emergency Planning Committee database, providing over 3,000 listings by state at http://www.epa.gov/ceppo/lepclist.htm#bystate.

What will the Emission Source Inventory database look like?

Figure 4-1 is an illustration of an Emission Source Inventory database with sample stationary point source data.

What happens once the Emission Source Inventory is completed?

The completed Emission Source Inventory database and summaries of the data should be presented at a meeting of the entire Partnership committee. The inventory can then be shared with the broader community, both for input and as an opportunity to educate the community on the project's progress. Once the database is complete, the work can then proceed to the next step, the Initial Screening step, where all the information collected in the Emission Source Inventory will be used to make the first cut at identifying the chemicals and sources that will be community priorities. This step is described in the next chapter.

What can the Partnership committee do to involve the broader community in the review of the Emission Source Inventory database?

The Partnership may want to prepare summaries and organize community meetings to ask the community to check to make sure that all of the local sources are included in the Emission Source Inventory database. Maps and the Geographic Information System (GIS) can be used to illustrate the sources and their locations. This will be an excellent opportunity to educate the community, both residents and businesses, on the sources and the releases that take place in and around the Partnership area. This will also be an excellent opportunity to build the Partnership and recruit new members from sectors of the community not yet involved in the work. Questions that are likely to arise concerning the meaning of this information and the possible effects of releases on community health will provide the Partnership with an opportunity to explain the goals of the Partnership and the next steps of the screening process.

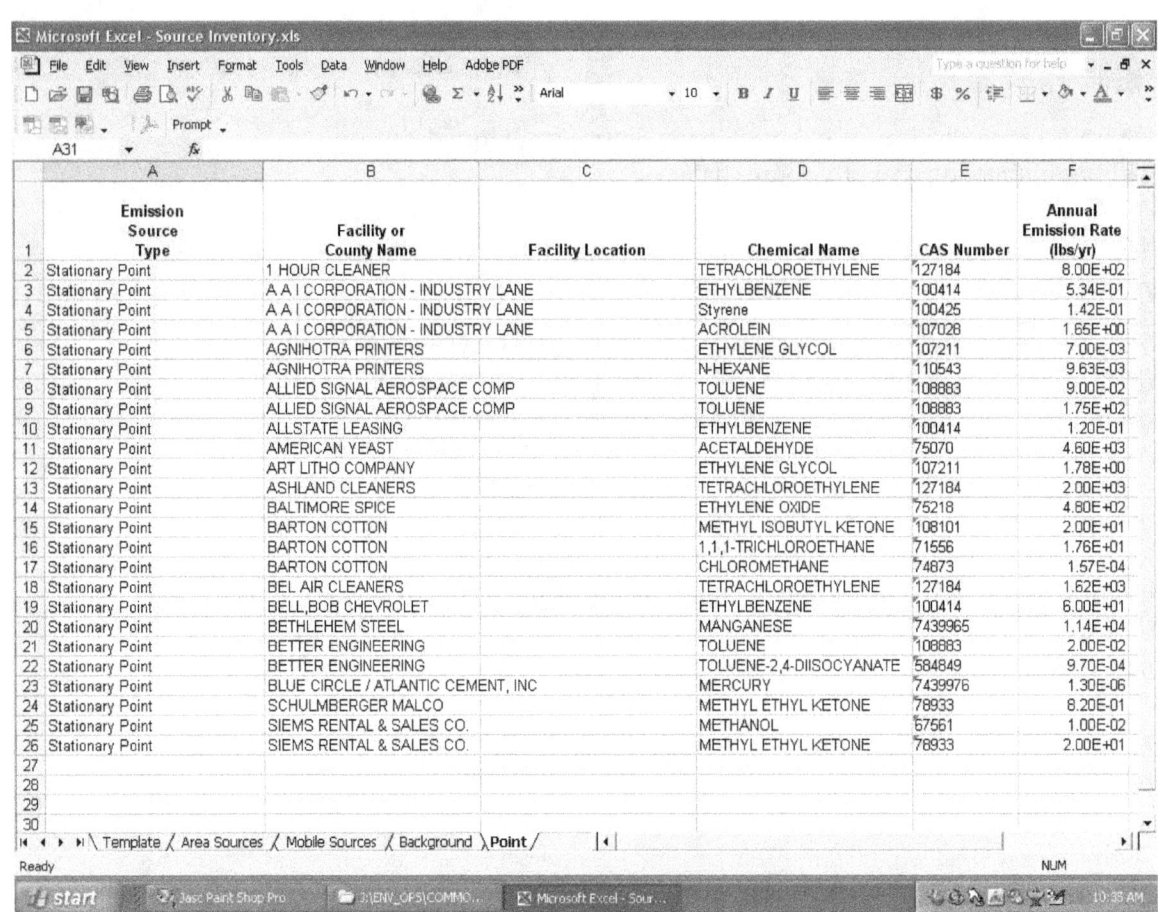

Figure 4-1.
An Illustration of an Emission Source Inventory Database

Initial Screen

Initial Screen Overview

Step 1: Estimating Concentrations

Step 2: Setting the Screening Level and Developing Screening-Level Concentrations

Step 3: Comparing Community Air Screening Concentrations to the Screening-Level Concentrations

Organize

Collect

Analyze

Mobilize

How is this explanation of the Initial Screen organized?

This Initial Screen section of the How-To Manual Overview is divided into the following four parts:

Initial Screen Overview: a general overview of the Initial Screen

Step 1: Estimating Concentrations: a description of the method that will be used to find or estimate concentrations of chemicals in community air

Step 2: Setting the Screening Level and Developing Screening-Level Concentrations: a description of how the Partnership will set its risk screening level and its risk-based screening concentrations

Step 3: Comparing Community Air Screening Concentrations to the Screening-Level Concentrations: a description of how the Partnership will compare the estimated concentrations to the screening-level concentrations to identify chemicals that need further review

Initial Screen Overview

How does the screening process work?

As described in the Introduction, the Initial Screen is the first of three screening steps designed to identify the chemicals in community air that have concentrations that are above the community risk screening levels. These three screening steps are designed to provide a practical way for a community to use the minimum possible amount of resources to sort through all the air releases in and around their area and identify the priority chemicals and sources. A general explanation of the screening methods used in this Manual is provided in Chapter 3. You may want to review that explanation as background for this chapter. A detailed explanation of the Initial Screen, the first of the three screening steps, is provided in this chapter.

What is the purpose of the Initial Screen?

With a large number of chemicals and sources, the task of accurately estimating concentrations in community air would be a resource-intensive effort for a community partnership. To avoid this problem, the Initial Screen uses a simpler method that requires fewer resources to review all the chemicals and sources to find those chemicals that have concentrations that are clearly at or below the community screening levels. Because these chemicals are at or below the community screening levels, they will be set aside as a lower priority. They will not need further review in this screening process. Eliminating these chemicals in the Initial Screen means that the resource-intensive methods needed to more accurately estimate concentrations that will be used in the Secondary and Final Screening steps will have a smaller number of chemicals to review.

Can the Partnership be sure that the chemicals eliminated from further review by the Initial Screen do not have concentrations in the community air that are above the community screening level?

The community can be confident that the chemicals eliminated during the Initial Screen do not have concentrations in community air that are above the community screening levels because the methods used to estimate concentrations in the Initial Screen are designed to overestimate the concentrations in community air.

Instead of using the resource-intensive methods necessary to more accurately estimate concentrations, the Initial Screen uses methods that assume a kind of "worst case" scenario that is designed to overestimate the concentrations. The method, for example, assumes that the direction of the wind in the community is always in the direction that results in the highest concentration and that the releases are always at the maximum permitted level. If the concentrations estimated using these worst-case assumptions are still at or below the community screening levels, then the community can confidently set these chemicals aside and focus on the chemicals that remain after completing the Initial Screen that may have concentrations above the community screening levels. (Also remember that all mobile source chemicals will be passed to the Secondary Screen for analysis.)

It is, however, important to keep in mind that the Initial Screen estimates are only as good as the information they are based on. If the releases are actually greater than the releases recorded in the Partnership's Emission Source Inventory database, or if the community inaccurately estimates the distance from a source to the nearest exposed community members, then the Initial Screen estimation of concentrations could be too low, i.e., the Initial Screen could underestimate instead of overestimate the concentration. Also, release amounts could be inaccurate if there are unknown sources or if the sources are releasing more than they are reporting. So confidence in the results of the Initial Screen depends on the accuracy of the information the Partnership has collected and entered in the Emission Source Inventory database.

It is also important to remember that chemicals set aside in this screening exercise should be reviewed periodically to see if there are any changes in release amounts or changes in any of the factors that might affect the screening levels. (These will be explained below.) Eliminating a chemical from further review in a screening exercise does not mean eliminating it permanently from review. All chemicals and releases should be reviewed periodically so that the Partnership can incorporate any new information that may affect the results of the screening. In addition, the Partnership may want to flag chemicals screened out during the Initial Screen that have concentrations that are close to the screening-level concentrations for further consideration in the future.

All chemicals and releases should be reviewed periodically so that the Partnership can incorporate any new information that may affect the results of the screening

How does the Initial Screen differ from the Secondary and Final Screens?

The Secondary and Final Screens differ from the Initial Screen only in the methods used for estimating concentrations. The level of risk and the screening-level concentrations developed in the Initial Screen will stay the same throughout the screening exercise. In place of the simple methods used to overestimate concentrations in the Initial Screen, the Secondary and Final Screening steps use more sophisticated methods and better release information to produce more accurate estimates for the concentrations of chemicals in community air.

What steps will the Partnership need to complete in the Initial Screen?

To complete the Initial Screening step, the Partnership will start with the information on releases collected and stored in the Emission Source Inventory database and use this information to complete the following steps.

Summary of Steps to Complete the Initial Screen

Step 1: Set up a quality assurance/quality control (QA/QC) procedure to ensure that all of the work of the Initial Screen is accurate and complete.

Step 2: For each chemical considered in the Initial Screen, estimate the ambient air concentration for all chemical emissions that results from all the sources of the chemical.

Step 3: Choose a level of risk to serve as the community's risk screening level.

Step 4: Calculate the risk screening concentration, i.e., the concentration equivalent to the risk screening level, for each chemical in community air.

Step 5: Compare the concentration for each chemical in community air to its risk screening concentration and identify the chemicals with concentrations above their screening levels. These chemicals will be looked at more closely in the remaining steps of the screening process.

A detailed description of each of these steps is provided in this chapter. Please note that mobile source chemicals and certain other sources will not be reviewed until the Secondary Screen.

How can the Partnership organize itself to effectively carry out the tasks needed to complete the Initial Screen?

To complete all the technical tasks necessary to estimate concentrations and to calculate community screening levels, the Partnership will probably need to appoint one or two teams from among its members. Partnership members with the necessary technical skills to complete the tasks will join with other interested members to complete the work and present it to the full Partnership committee for review and approval. If enough resources are available, organizing two teams, one for estimating concentrations and one for developing screening concentrations, will save time, since these tasks can move forward simultaneously.

St. Louis Community Air Project (CAP) team at work.

What are the key decisions that the Partnership will need to make during the Initial Screening to ensure that screening meets its needs?

There are two important decisions that will need to be made during the Initial Screen that will require input from the full Partnership committee. First, the Partnership will need to decide on the level of risk to use for screening. Setting the level of risk that the community agrees to use as the screening level is a key decision for the Partnership and the community. A full discussion of the meaning of this risk screening level and a discussion of factors to consider in setting this level are provided later in this chapter.

There are two important decisions that will need to be made during the Initial Screen that will require input from the full Partnership committee

To make the decisions needed to calculate screening concentration values, the Partnership will need to consider a broad range of views and perspectives

Second, the full Partnership committee will need to make decisions to ensure that the screening-level concentrations that it will use to screen chemicals meet its needs. As explained below, the full Partnership committee will rely on a technical team from among its members to calculate screening concentrations. To ensure that its technical team calculates screening concentrations that meet the Partnership's needs, the full Partnership committee will have to decide on the rules and standards that its technical team will follow to develop the screening-level concentrations. The participation of the full Partnership committee in choosing the level of risk and setting the rules for the development of the screening concentrations will ensure that the screening steps are carried out in a way that best reflects community values and needs.

What new information will be needed to complete the Initial Screen?

The Partnership's Inventory Team will collect the information about emissions needed to estimate ambient air concentrations during the Initial Screen. The new emissions information that will be needed is described below. The Partnership will also need to collect the information needed to calculate the screening-level concentrations for each of the chemicals in its Emission Source Inventory database. These screening-level concentrations will be used during the Initial, Secondary, and Final Screening steps. The Partnership team with the responsibility for calculating screening-level concentrations will collect this information and add it to the Emission Source Inventory database. This information is also described below. Detailed guidance to help the teams collect this information can be found in Chapter 10, in the Technical Guidance section of the Manual.

What resources and skills will be needed to complete the Initial Screen?

The Initial Screen will require the skills and resources of most of the Partnership committee. The following is a list of the skills and resources the Partnership will need to complete the Initial Screen.

Deliberative skills: In its decisions on the level of risk to use for the screening process and on the standards for choosing and calculating screening concentration values, the Partnership will need to consider a broad range of views and perspectives. The ability of the members of the Partnership to use deliberative skills to consider different viewpoints and help the Partnership find common ground on this key issue will be crucial to the success of the project. It will also be important for the Partnership to solicit input from the broader community on these decisions so that the community is prepared to support potential actions that might be needed to improve local air quality.

Technical skills: The Partnership committee will need to appoint teams from its members to carry out the technical work of estimating concentrations and finding or calculating screening values. To accomplish these tasks for the Partnership, the technical teams will need expertise with air dispersion modeling, risk, and the calculation of risk screening concentrations. Familiarity with available databases with screening values and toxicity data would also be helpful. Science teachers, science students, science faculty, technical government staff, retired engineers and scientists, and industry scientists could provide these skills. Expertise in the science of toxicology will be especially important for calculating screening-level concentrations.

Science education skills: Most of the risk science that will be used in the screening process is introduced in the Initial Screening step. It will be important for all the members of the Partnership committee to understand and be comfortable with this science. Individuals with skills in explaining science and technical models to non-technical audiences will be needed to help with the explanation of this science and its use in the screening process. Teachers and technical writers on the committee could provide these skills.

Communication skills: Explaining the Initial Screen and its results to the broader community will be an important part of the Partnership's work during and at the completion of the Initial Screen. (See the discussion of communication in Chapter 2.) The Partnership committee will need individuals with the ability to communicate the work of the Partnership to the community. Community members, teachers, journalists, or possibly a team that combines community members with professional writers could provide these skills.

How can the Partnership carry out the work of the Initial Screen in a way that helps to build the long-term capacity of the community to understand and improve air quality?

Here are a couple of ideas that may help the Partnership to conduct its work in a way that builds the long-term capacity of the community to understand and improve air quality. First, making the decision on the level of risk that will be used for screening, a key Partnership and community decision, will require considerable discussion, debate, and education. If the Partnership conducts these discussions and makes its decision in a way that respects and values the opinions of all of its members, the experience will go a long way to building trust and strengthening the relationships among the Partnership members. A strong and effective working relationship among all the members of the Partnership is an essential component of long-term community capacity.

A strong and effective working relationship among all members of the Partnership is an essential component of long-term community capacity

Second, the Initial Screen presents an excellent opportunity to provide the background education that the broader community will need to understand and participate in the work to improve local air quality. Providing education and organizing discussion in the community focused on the key decisions associated with setting and calculating risk screening levels will help to develop the community's long-term capacity to address air quality issues.

And third, working with community residents, teachers, and students to estimate concentrations and calculate screening levels will help to ensure that the technical skills needed for future work exist within the community. The technical teams could focus on training a core of community residents and teachers in the skills necessary to complete the technical work of the Initial Screen and then work with them to carry out the work. Establishing a center for this work in a local high school or college would also help to ensure that the community maintains these skills. Focusing on using the technical work to build the long-term capacity of the community will take more resources, but the benefits gained in building community capacity and community confidence in the validity of the work may justify the extra time and effort.

Step 1:
Estimating Concentrations

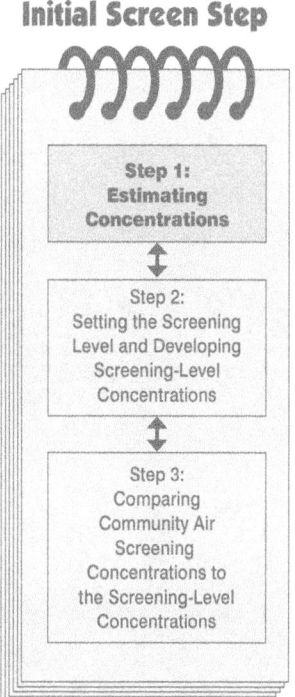

Where does the information on air concentrations of chemicals come from?

There are two sources for the information on the concentrations of chemicals in community air. The first source is the ambient air monitoring stations that may be located in or close to the Partnership area. These monitoring stations collect air samples on a regular basis. These samples are analyzed to determine the concentration of chemicals present in the air at the monitoring station. The measured concentrations can be combined to provide annual average concentrations for the chemicals that are monitored. Local monitoring data is a very important resource and it is included in the Emission Source Inventory database prior to any modeling exercise to help identify potential chemicals of concern that could require additional attention.

While these monitored concentrations are an accurate measure of concentrations at specific locations, the number of monitoring stations is limited, and many communities and neighborhoods may not have a station in or near their area. In addition, monitoring stations measure concentrations at the station locations only, so, depending on the location of sources in the community, concentrations at other locations in the community may be different from the concentrations at the monitoring station.

The second source for information on the concentration of chemicals in community air, the primary source used in this Manual, comes from air dispersion modeling. Air dispersion modeling uses a computer to combine information on local wind direction and meteorological conditions and the behavior of chemical plumes to model the dispersion of chemicals as they are released from different sources. (Air dispersion modeling is described in more detail below.) While air dispersion modeling relies on modeled estimates and not on actual measurements taken at monitoring stations, this method of estimating concentrations has been extensively tested and verified to provide reliable estimates of the concentrations that result from chemical releases. Information on the reliability of air dispersion modeling is available in Appendix W at the Support Center for Regulatory Air Models (SCRAM) web site at http://www.epa.gov/scram001/tt25.htm#guidance.

Air dispersion modeling can also provide some information that monitoring cannot. With air dispersion modeling, a community has the ability to estimate concentrations at key locations in the community where no monitors are located, such as at a school or near a group of homes. Air dispersion modeling can also be used to find the location in a community with the highest concentration of a chemical. And, perhaps most importantly, air dispersion modeling makes it possible for a community to estimate the contribution that different sources make to the concentrations identified as community priorities. This provides the information needed to target the effort to reduce releases from selected sources to the appropriate levels.

To use air dispersion modeling, it is important to remember that modeled estimates are only as good as the information that is put into the model. Accurate estimates of air concentrations depend on good information about the sources and their releases. This information is collected by the Partnership and stored in its Emission Source Inventory database.

How does air dispersion modeling work?

As chemicals are released from a source, they form a plume and begin to disperse. The smoke plumes that we commonly see leaving chimneys or smokestacks are examples of visible plumes of pollutants dispersing in the air. Scientists have studied these plumes and developed models that can predict how chemicals will disperse as they leave a source and then can estimate the resulting concentrations in surrounding locations. These models are called air dispersion models, and many of them are available for use on desktop computers.

An example of a plume. Air dispersion modeling can be used to estimate concentrations of chemicals resulting from these and other types of releases.

What air dispersion models will be used in the screening process?

The screening process in the How-To Manual utilizes two EPA air dispersion models: SCREEN3 for the Initial Screen and ISCST3 for the Secondary and Final Screens. For the air dispersion models to estimate concentrations, information about the sources and their releases must be collected and entered into the models.

How will the SCREEN3 air dispersion model be used in the Initial Screen?

With all the sources and releases affecting community air to review, the Partnership needs a simple way to use air dispersion modeling to estimate concentrations. These concentrations will be reviewed during the Initial Screen to identify those that will need more detailed analysis during the Secondary and Final Screens and those that can be eliminated because they are definitely below the community screening levels. Setting up and running an air dispersion model for all the chemicals and sources in the Partnership's Emission Source Inventory database would require a lot of resources, so for the Initial Screen, the SCREEN3 air dispersion model was used to develop a look-up table that the Partnership can use to quickly estimate ambient air concentrations for many of the releases in its database. This look-up table was developed by running the SCREEN3 model for a unit release amount for a range of different stack heights and receptor distances. The results from SCREEN3 modeling were used to create the look-up table shown on Table 5-1. (The mathematical shorthand used in this table will probably be unfamiliar to most non-technical members of the Partnership. Members of your technical team will be able to explain it and demonstrate that using the table to estimate concentrations is actually quite easy.)

To use this look-up table to estimate a concentration for stationary sources, the Partnership's technical team will use the information on stack height and the distance to the closest receptor in the Emission Source Inventory database to find the appropriate unit concentration in the look-up table. The team will use this unit concentration and the release amount to calculate an estimate of the concentration resulting from the release. This procedure will be followed for each of the stationary source releases in the Emission Source Inventory database. If the technical team and Partnership do not have sufficient resources to collect the information needed to use the look-up table, an alternative method of estimating ambient air concentration values using a simple conversion factor based on the SCREEN3 look-up table can be used to estimate concentrations. A description of this conversion factor and discussion of its potential use can be found in Chapter 10 of this Manual. This conversion factor is also used to estimate the concentrations resulting from stationary area sources. Chapter 10 also provides guidance to help the technical team estimate concentrations resulting from releases from stationary area sources.

Table 5-1. Look-Up Table.
Unitized Annual Average Concentration at Different Distances from a Stack ($\mu g/m^3$ per lb/year)

Stack Height (m)	Distance to Receptor							
	15 m	20 m	50 m	100 m	200 m	300 m	400 m	500 m
3	1.59E-02	1.55E-02	1.04E-02	3.97E-03	1.19E-03	5.78E-04	3.50E-04	2.39E-04
5	5.63E-03	7.05E-03	5.42E-03	3.33E-03	1.13E-03	5.64E-04	3.45E-04	2.37E-04
8	1.05E-03	2.20E-03	2.93E-03	2.23E-03	1.01E-03	5.33E-04	3.33E-04	2.31E-04
10	2.35E-04	9.51E-04	1.94E-03	1.57E-03	9.12E-04	5.08E-04	3.23E-04	2.26E-04
20	1.70E-09	1.08E-06	4.05E-04	4.59E-04	3.37E-04	2.79E-04	2.06E-04	1.55E-04
30	-	-	6.98E-05	2.10E-04	1.76E-04	1.30E-04	1.22E-04	1.04E-04
40	-	-	6.34E-06	1.02E-04	1.02E-04	8.53E-05	6.49E-05	6.49E-05
50	-	-	3.00E-07	4.79E-05	7.03E-05	6.08E-05	4.94E-05	3.81E-05
75	-	-	-	3.74E-06	2.95E-05	2.95E-05	2.54E-05	2.36E-05
100	-	-	-	1.11E-07	1.28E-05	1.63E-05	1.59E-05	1.41E-05

Why are these simplified methods for estimating concentrations using a look-up table appropriate for the Initial Screen?

As discussed in the Overview section of this chapter, the simplified use of air dispersion modeling, incorporated in the look-up table, assumes that the direction of the wind in the community is always in the direction that results in the highest concentration. This assumption, and the method for combining concentrations from different sources discussed below, means that concentrations estimated using the look-up table are very likely to be overestimated, i.e., the estimated concentrations will likely be higher than the actual concentrations in the community air. The fact that the look-up table overestimates concentrations makes it appropriate for use as the first step in this screening process. If the concentrations estimated using the look-up table are below the community screening values, then a more accurate estimate of the concentrations would be even further below the screening values. As a result, the look-up table based on the SCREEN3 model can be used to identify and eliminate those chemical releases that need no further review. Limiting the number of chemicals in the review process will make the resource-intensive modeling in the next steps of the screening more manageable. That, in fact, is the purpose of this Initial Screen: to provide an easy way to find and eliminate chemicals with concentrations that are definitely below the community screening levels to minimize the resources that will be needed to perform the more sophisticated air modeling required during the Secondary Screen.

Who will the Partnership rely on to complete the technical work of estimating concentrations in the Initial Screen?

The Partnership will need to appoint a technical team from among its members to estimate the concentrations for each chemical in the Emission Source Inventory database. Detailed guidance to help this team estimate concentrations is provided in the first section of Chapter 10. The technical team will need to establish clear quality control procedures to ensure that the estimated concentrations are calculated and recorded accurately for the Partnership. When the team has completed its work, it can present the results to the full Partnership committee for review. The Partnership's Inventory Team will work with the technical team estimating concentrations to collect the additional information that will be needed to complete the Initial Screen. The information requirements for the Initial Screen are discussed below.

What skills will the technical team need to estimate concentrations in the Initial Screen?

Skills needed will include:

- Ability to create, import, manipulate, and query databases
- Experience in using air dispersion modeling to estimate concentrations
- Ability to implement QA/QC procedures

College and university students and faculty, high school teachers and their students, interested community residents, retired professionals, government staff, and industry and business technical staff are all likely to have the necessary skills.

Which of the sources and chemicals will have their concentrations estimated using the SCREEN3 look-up table?

The Emission Source Inventory database contains information on four types of sources: stationary point, stationary area, mobile on-road, and mobile non-road. Stationary point sources include all large and small commercial, industrial, and publicly owned facility sources. Stationary area sources include household sources such as heating, wood stoves, and other kinds of sources that are too numerous or difficult to estimate individually. Mobile on-road sources include all trucks, buses, cars, and any other street or highway vehicles. Mobile non-road sources include trains, airplanes, ships, and construction and lawn equipment. The SCREEN3 look-up table and conversion factor based on the look-up table will be used to estimate the ambient air concentrations for **stationary point and stationary area sources** only. Concentrations resulting from **mobile sources will not be estimated** using the look-up table. This look-up table was developed to estimate concentrations for stationary sources only. There is no simple way to use air dispersion modeling to develop a similar look-up table for mobile sources. In addition, concentrations resulting from mobile source emissions are well documented and are likely to be close to or above the community screening levels. (A list of the mobile source chemicals is provided in Table 10-4.) For these reasons, all the chemicals associated with mobile sources will be automatically moved to the Secondary Screen, where more sophisticated air dispersion modeling will be used to estimate their concentrations. This also means that if any chemicals released from stationary sources are also mobile source chemicals, the Partnership will not estimate their concentrations during the Initial Screen. Instead, the stationary source releases of these chemicals will be evaluated during the Secondary Screen.

In addition, releases from all stationary point sources that are less than 15 meters (approximately 49 feet) from a place where community members might be exposed, such as a home or school or business, will also be automatically moved from the Initial Screen to the Secondary Screen. The look-up table is not designed to screen for concentrations at these close distances, so to ensure that these sources are adequately reviewed, their concentrations will be estimated using the more sophisticated modeling in the Secondary Screen. The maximum distance to receptor in the look-up table is 500 meters. Receptors that are located further than 500 meters from the source are modeled at 500 meters using the look-up table. This is conservative because concentrations decrease as the distance to the receptor increases.

Here is a summary of the sources that will and will not be reviewed in the Initial Screen.

Sources Reviewed in Initial Screen	Sources Not Reviewed in Initial Screen
• Stationary point sources more than 15 meters from closest community exposure	• Mobile sources, on-road and non-road
• Stationary area sources	• Stationary point sources less than 15 meters from closest community exposure

What information will be needed to use the SCREEN3 air dispersion model?

The information for each source that will be needed to use the SCREEN3 model includes the following:

For stationary point sources:

- Amount of release (pound/year) for each chemical
- Type of release: stack or fugitive
- Approximate height of the stack (meters)
- Distance to the closest exposed person in the community (meters)

For stationary area sources:

- Total county release amount of each chemical from each area source (pound/year)

Stationary point source releases are divided into stack and fugitive release types in the Emission Source Inventory database. Examples of stack releases include a release from a roof vent of a dry-cleaning business and a release from a large stack like the kind used for most municipal trash incinerators. Examples of fugitive releases include emissions from the surface of a treatment pond or from the nozzles at a gas filling station.

All the information needed to complete the Initial Screen will be collected and stored in the Partnership's Emission Source Inventory database by the Inventory Team. An explanation of how this information will be used in the Initial Screen to estimate concentrations is provided in the next section.

How will the Partnership's technical team estimate concentrations for the stationary point sources?

Detailed guidance and examples for estimating concentrations during the Initial Screen are provided for the technical team in Chapter 10. The overview provided in this and the next sections are designed to provide all the Partnership members with an understanding of the work that will be done by its technical team.

Concentrations from stack and fugitive releases from **stationary point sources** will be estimated as follows:

- To estimate a concentration for each chemical resulting from a stack release, the technical team will use the information contained in the Emission Source Inventory database on release amount, release type, stack height, and the distance to the closest community exposure with the look-up table to calculate the concentration resulting from the stack release. The estimated concentrations resulting from the stack releases from all the stationary point sources 15 or more meters from the closest community exposure will be calculated.

- To estimate a concentration for each chemical resulting from a fugitive release, the technical team will use the same look-up table and assume that the release is from a three-meter stack. Although the look-up table is designed for stack releases, using this assumption will ensure that the concentrations resulting from the stationary fugitive source releases are overestimated, as is appropriate for the Initial Screen.

When completed, the technical team will have a list for each chemical of the concentrations resulting from the stationary point sources, stack and fugitive, releasing the chemical. (If resources are limited, the technical team may choose, with Partnership consent, to eliminate the collection of new information and use a simpler conversion factor, discussed in Chapter 10, instead of the look-up table to estimate concentrations.)

How will the technical team estimate the concentrations from the stationary area sources for the Initial Screen?

Stationary area sources include household sources such as heating, wood stoves, and other kinds of sources too numerous or difficult to estimate individually. To estimate the concentrations resulting from these sources, the technical team will, for each chemical, add the county-wide release amounts from all the area sources together and then estimate the concentration by using a conversion factor based on SCREEN3. A detailed description of this conversion factor can be found in Chapter 10. Using this conversion factor and the county-wide releases for each source, instead of the releases that come from the sources in the Partnership area only, will likely result in overestimation of the concentrations appropriate for the Initial Screen.

How will the concentrations that result from the combined releases of the same chemical from all the sources in the community be estimated?

If there are multiple stationary sources of the same chemical in or around the Partnership area, the concentration in air at a location in the community may be the result of a combination of these sources. This occurs when the dispersion plumes from the sources overlap. Since the look-up table only provides estimates for each point source and for the area sources separately, some method for estimating the combined (aggregate) concentration resulting from all the stationary sources must be included in the Initial Screen.

The simplest method to accomplish this is to add the concentrations from all the stationary sources. This is very likely to be an overestimate since not all the plumes will overlap, especially those from small sources. And where the plumes from sources do overlap, the concentrations at the point of overlap will not always be the highest concentration, which is the concentration that is calculated using the look-up table. But, again, overestimation is appropriate during the Initial Screen since the goal is to eliminate chemicals that are clearly below the community screening values.

Finally, to estimate the total concentration resulting from all sources reviewed in the Initial Screen, the technical team will look in the Emission Source Inventory database to see if there are any background concentrations for the chemicals under review. These background concentrations will then be added to the total stationary source concentrations to get the total estimated concentration for a chemical. This is the concentration that will be compared to the community's screening concentration for that chemical.

What are all the steps the technical teams will need to complete to estimate the concentrations for each of the chemicals released by stationary sources?

The following page has a summary of all the steps the Partnership's technical teams (the Inventory Team and the Concentrations Estimation Team) will need to complete to estimate concentrations in the Initial Screen, starting with all the chemicals in the Partnership's Emission Source Inventory database.

Summary of Steps to Estimate Concentrations in the Initial Screen

Step 1: Set up QA/QC procedures for data collection, data entry, and calculations.

Step 2: Collect information on release type, stack height, and receptor distance for stationary point sources. (If the technical team has limited resources and chooses to use the conversion factor to estimate concentrations, then skip this part of step 2.) Check to be sure that county-wide release amounts for all stationary area sources have been collected.

Step 3: Identify all stationary point sources that are less than 15 meters from the nearest community exposure and move all the chemicals emitted from these sources to the Secondary Screen for evaluation. These chemicals will not be reviewed during the Initial Screen.

Step 4: Move all the chemicals from mobile sources to the Secondary Screen for evaluation. These chemicals will not be reviewed in the Initial Screen. (Note: In urban areas this means that many important monitored pollutants will not be examined during the Initial Screen.)

Step 5: For each chemical and for each of the stationary point sources releasing the chemical, estimate the concentration using the look-up table or conversion factor method. Add all the concentrations to get a total point source concentration.

Step 6: For each chemical released from stationary area sources, add the county-wide releases from all the area sources and estimate the concentration using the conversion factor.

Step 7: Obtain the background concentration, if any, for each chemical.

Step 8: For each chemical, add the total stationary point source concentration, the total stationary area source concentration, and the background concentration, to obtain the total concentration from all sources.

What are the next steps once the technical team has estimated concentrations for all the chemicals in the Emission Source Inventory database?

Once the Partnership has estimates for the concentrations of chemicals from stationary sources in the community's air, it will compare these concentrations to the Partnership screening-level concentrations. Any chemical with a concentration above the Partnership screening level will be kept for further review. All other chemicals will be removed from the screening process and set aside as low priority for now. As mentioned above, these chemicals should be reviewed periodically to check for any changes, such as increased releases or new toxicity information, that might change the decision to drop the chemical from the review process.

Step 2: Setting the Screening Level and Developing Screening-Level Concentrations

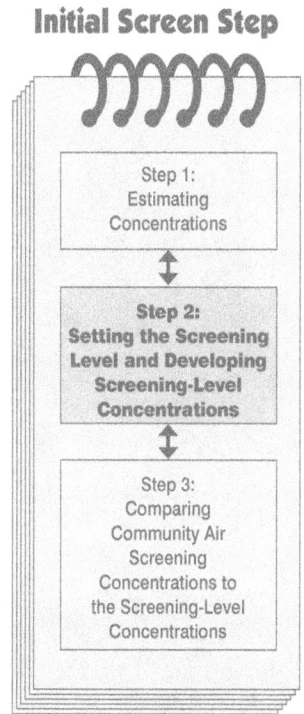

Initial Screen Step

Step 1: Estimating Concentrations

Step 2: Setting the Screening Level and Developing Screening-Level Concentrations

Step 3: Comparing Community Air Screening Concentrations to the Screening-Level Concentrations

Now that the Partnership has estimated the concentrations in community air for the chemicals in its inventory, how will it begin the screening process to identify the priority chemicals that will be evaluated for possible community action?

To begin the screening process, the Partnership chooses the risk screening level it will use to identify priorities. Then, using this risk screening level, the Partnership calculates screening-level concentrations for each chemical. And, finally, the Partnership compares the concentrations in community air that were estimated using the method discussed in the previous section to the screening-level concentrations. The chemicals with concentrations in community air that are above the screening-level concentrations will be kept for further review. Chemicals with concentrations that are at or below the screening-level concentrations will be set aside. (Please refer to the more detailed explanations of the screening method in Chapter 3.)

The following is an illustration of the use of the full screening process to identify a priority chemical for the community.

A cancer-causing chemical, such as benzene, is one of the chemicals present in community air. The Partnership chooses a one in 1 million risk level as its risk-screening level. It then calculates the lifetime continuous exposure concentration for benzene associated with that risk level. This concentration becomes the Partnership's benzene screening level. The Partnership compares its estimated ambient air concentration for benzene to the screening-level concentration for benzene. If the estimated benzene concentration in community air is above the benzene screening-level concentration in the Initial, Secondary, and Final Screens, the Partnership assumes the risk associated with benzene's estimated concentration would be more than one in 1 million and, as a result, the Partnership identifies benzene as a priority chemical to be evaluated for possible community action.

The detailed explanation of and guidance for choosing the community risk screening level and calculating the screening level concentrations equivalent to the risk screening level are provided in this section and in Chapter 10 of the Technical Guidance section of the Manual.

Please note that there is an alternate approach to risk-based screening that uses a proportion of risk and hazard instead of screening-level concentrations to screen chemicals. While this alternate approach is not detailed in this Manual, a brief description of this method is provided later in this chapter.

What skills and resources will the Partnership need in order to find or calculate the screening-level concentrations?

While the screening exercise will require input from all the members of the Partnership, the work to develop the screening-level concentrations will require special technical skills. As a result, the Partnership committee will need to appoint a technical team from among its members to develop the screening-level concentrations that the Partnership will use. Technical guidance to help this technical team can be found in Chapter 10 of this Manual.

To develop screening-level concentrations, the technical team will need the following skills and resources:

- Working knowledge in one or more scientific fields, including chemistry, biology, toxicology, industrial hygiene, and public health. These experts will also need to identify and use the types of numerical (quantitative) information used to calculate screening-level concentrations.

- Computer skills to access and search web sites for the most current and best available chemical toxicity information for calculation of screening-level concentrations.

- Familiarity with the use and management of computerized databases (e.g., those that provide existing screening-level concentrations and chemical toxicity information). Such familiarity would be helpful for identification, retrieval, and storage of information used to calculate screening-level concentrations.

- Computer skills for the calculation, storage, and presentation of community screening-level concentrations.

Professionals living or working in the community, health professionals, college and university faculty and students, high school science teachers, government, industry and business technical staff, and retired professionals can provide these skills.

What is a community risk screening level?

Before the technical team can calculate screening-level concentrations, the Partnership will need to choose the level of risk that will serve as the community's risk screening level. Each Partnership and community will develop its own risk screening level, based on its own goals and values. The risk screening level will be used to identify community priorities. By agreement of the Partnership, any chemical in community air that results in a level of risk above the risk screening level will become a community priority and will be evaluated for possible community action. These priority chemicals will be identified during the Final Screen at the end of the screening process.

Once agreed upon, the risk screening level will be used to develop the screening-level concentrations for all of the community chemicals. The method for developing screening-level concentrations will be discussed below.

The Partnership risk screening level relates specifically to risks of adverse health effects caused by long-term (chronic) exposure to chemicals in the community. Short-term (acute) effects are not considered in this screening process. Screening levels for chronic exposures are generally lower than acute screening levels. Acute exposures were not included in the Baltimore project, so step-by-step guidance for including them in a screening process are not included in this Manual. In particular, guidance for collecting information on the peak releases and meteorological conditions that may result in high short-term concentrations, guidance for estimating short-term concentrations, as well as guidance for developing the screening-level concentrations for acute effects are not included. Acute effects are likely to be important community concerns, so we recommend that the Partnership, if possible, draw on it technical members to develop an effective way to identify and address these concerns.

Are there different risk screening levels for cancer and non-cancer effects?

For long-term chronic exposures, there may be two types of risk screening levels from which the Partnership will calculate screening-level concentrations, depending on whether or not the chemical is considered carcinogenic. The Partnership will use one type of risk screening level for cancer and a second type of risk screening level for other effects. When a chemical has a potential to cause cancer as well as some other long-term health effect, the Partnership will calculate two screening-level concentrations for the chemical, using each of the two types of risk screening levels. The Partnership will then use the screening-level concentration that provides the most conservative or most protective (i.e., the lowest) concentration as its screening level concentration. In most cases, this will be the screening-level concentration calculated by using the Partnership's cancer risk screening level.

The risk screening level used to calculate a screening-level concentration when cancer is the long-term toxicity concern is expressed in statistical terms called a unit risk. A unit risk is a measure of a probability of risk that the Partnership considers appropriate for identifying priorities for chemicals with some identified cancer-causing potential. For example, a community's priority-setting level of risk for these potentially cancer-causing chemicals might be set at a risk of one excess cancer case in 1 million exposed people. This means that if a chemical in community air has a concentration level that is estimated to cause more than one in 1 million excess cancer cases, then it will be identified as a priority chemical and will be evaluated for possible community action.

As is discussed below, several EPA programs have used one in a million excess cancer cases as a risk screening level. The Partnership can select a more or less conservative target cancer risk level. For example, the Partnership could choose the more conservative risk of one excess cancer case in 10 million exposed people, or the less conservative risk of one case in 100,000 exposed people. Whatever value the Partnership agrees on as its cancer risk screening level, it remains the same for the calculation of the screening-level concentration for each chemical with some identified cancer-causing potential.

The risk screening level that is used to calculate a screening level concentration for non-cancer effects is expressed as a ratio. This ratio is referred to as a *hazard quotient,* and it is used by the Partnership to develop screening concentrations for non-cancer effects. The hazard quotient is the ratio of the potential exposure to the substance and the level at which no adverse effects are expected. If the hazard quotient is calculated to be less than 1, then no adverse health effects are expected as a result of exposure. If the hazard quotient is greater than 1, then adverse health effects are possible. Unlike the cancer risk screening level expression, the hazard quotient cannot be translated to a probability that adverse health effects will occur, and is unlikely to be proportional to risk. It is important to note that a hazard quotient exceeding 1 does not necessarily mean that adverse effects will occur.

For example, a community's level of risk for non-cancer-causing chemicals might be expressed as a hazard quotient of 1. In this example, the concentration in community air that would identify a chemical as a community priority would be a concentration that is greater than the concentration used as a measure of the chemical's toxicity potential. A hazard quotient of 1 is used by several EPA programs as a trigger for considering action. Just as in the case of cancer-causing chemicals, the Partnership can choose a risk screening level more or less conservative than a hazard quotient equal to 1 for its non-cancer risk level. And, as for the cancer-causing chemicals, whatever value the Partnership agrees on as its non-cancer risk screening level, it will remain the same for calculating the screening-level concentrations for each chemical that can cause long-term, non-cancer effects.

What should the Partnership consider in choosing a risk screening level to use for its screening-level concentration-setting process?

The Partnership needs to choose a risk screening level that fits community values and helps to identify priorities, so that local air quality can be effectively improved. There are several considerations that may help a Partnership choose a risk screening level that meets these two goals.

First, it may be helpful to remember that the risk screening level is the level of risk that a community agrees should be used to identify the priority chemicals and sources that will be evaluated for possible community action, most likely voluntary, to reduce risk. Since the risk screening level is based on a community agreement, it can only result from a dialogue and an effort to build a consensus within the community. The broader the consensus, the more meaningful the screening level will be. Taking the time to build a community consensus is essential to the effort to improve air quality.

Taking the time to build a community consensus on a risk screening level is essential to the effort to improve air quality

Also, keep in mind that setting the risk screening level does not bind a partnership to any kind of specific action. The Partnership will need to carefully evaluate the options that are available for addressing priority chemicals and sources before deciding on the course of action that best meets community needs. (See Chapter 8 for suggestions on developing options for addressing priority chemicals and sources.) Feasibility and cost will need to be considered and balanced with other community concerns to find the best course of action. Some sources identified as community priorities may require an action, such as a vehicle inspection program, that is beyond the scope of any single local community. In that case, the action identified by the community may be an effort to influence other communities to join in the larger effort that would be required to initiate a regional or even national risk reduction effort. Efforts to reduce emissions from other kinds of sources, such as local businesses, households, and local diesel-powered truck and bus traffic, can be more easily addressed by a local Partnership alone. So, the risk screening level will help to identify priority air risks, but it cannot determine how, or if, the Partnership will decide to address them. Only careful consideration by the Partnership will determine how to best address the priorities identified by the screening level.

In choosing a risk screening level, it is also important to remember that it can be either too stringent or not stringent enough to effectively identify priorities. If a Partnership sets a very conservative (stringent) screening level, too many of the chemicals in community air could be identified as priorities. Unless the Partnership devises some way to prioritize this list of chemicals, it will not be able to find the chemicals that have the greatest potential to adversely affect the health of members of the community. On the other hand, setting a risk screening level that is not stringent enough could result in too few or no chemicals being identified as priorities. This would also make it difficult for a Partnership to identify the actions that would most effectively reduce air risks in the community. Communities can experiment with different target levels of risk to find the one that results in the most effective actions to improve local air quality.

It is also important to remember that the real goal of everyone in the Partnership is to improve local air quality. The decision to set a target level of risk should not take so long that it delays a community from getting to the work to improve local air quality. Setting a risk level that everyone can agree on, conducting the screen, and finding and getting to work on practical efforts to reduce risk may be more important than trying to decide, finally, on what is the best screening level for the Partnership to use. Once work is under way, where feasible, on the chemicals that have the greatest potential to affect the health of members of the community, the Partnership can return to the question of the risk screening level, adjust it, and identify more chemicals to be evaluated for possible risk reduction.

Finally, the Partnership may want to review examples of risk screening levels that have been used by other communities and by governments for different environmental programs. A few examples are provided below.

Baltimore Community Environmental Partnership: The Partnership in Baltimore chose a risk screening level of one in 1 million for cancer-causing chemicals, and a hazard quotient of 1 for non-cancer-causing chemicals. There was considerable discussion in the Baltimore Partnership over the choice of this screening level, especially since the state of Maryland uses a risk screening level of one in 100,000 at the fenceline of a facility. The Partnership decided to use the more conservative screening level and make a special effort to make sure that the community understood that any actions taken based on its screening level would be voluntary. The screen resulted in identifying four priority chemicals to be evaluated for possible community action. Details of the Baltimore screening exercise can be found in the Baltimore Case Study. This study is available at http://www.epa.gov/oppt/cahp/case.html.

St. Louis Community Air Project (CAP): The St. Louis screening project chose one in 100,000 as their cancer risk screening level, partly because they are using air monitoring instead of modeling to measure concentrations, and, for some of the chemicals, the monitors are not able to accurately measure the very small concentrations associated with the one in 1 million risk level. The educational handout prepared for the St. Louis Partnership's discussion of target risk levels is included in Appendix F.

What is a community screening-level concentration?

Once the Partnership has chosen a risk screening level to use for screening, the next step is to calculate the air concentration that corresponds to the community's risk screening level for each chemical in community air. In other words, the Partnership must calculate for each chemical the concentration that would result, after a lifetime of exposure, in a risk equal to the risk screening level chosen by the Partnership. These concentrations will be used by the Partnership as the screening-level concentrations. The concentrations in community air estimated during the Initial Screen will be compared to these screening-level concentrations.

To understand the use of screening level concentrations, it may be helpful to look at another benzene example. If the Partnership chooses one in 1 million excess cancer cases as its risk screening level, and if benzene is found in community air, the Partnership will need to determine the concentration (the level or amount) of benzene in the air that will result in a risk of one in 1 million in a hypothetical person continuously exposed to that concentration for a lifetime. This will be the screening-level concentration for benzene. The Partnership will compare the benzene concentration in community air that it has estimated to the benzene screening-level concentration. If the benzene concentration in community air is larger than the benzene screening level concentration, the risk from the benzene concentration is above the risk screening level, one in 1 million, and benzene would be kept for further review.

The Partnership will need to find or calculate a risk screening level concentration like the benzene example used above for each chemical in community air. The risk screening concentration for each chemical will be different because the toxicity of each chemical is different. For example, if you compare the screening-level concentrations for two chemicals with different toxicity, the screening-level concentration for the more toxic chemical will be smaller than the screening level for the less toxic chemical. This is because it takes a relatively small concentration of a more toxic chemical to produce a risk equivalent to the risk screening level. The following examples of risk screening values for hexavalent chromium, benzene, and toluene used by the Baltimore Partnership illustrate the difference in concentrations that result in an equivalent one in 1 million risk, the risk screening value chosen for use by the Baltimore Partnership. (These screening concentrations may differ from those calculated by your technical team for use in your screening exercise, since the method recommended in this Manual for calculating screening concentrations

Example screening-level concentrations used in Baltimore study	
Hexavalent Chromium	0.00015 µg/m³
Benzene	0.22 µg/m³
Toluene	420 µg/m³

is different from the method used in the Baltimore project. See the discussion below.)

Remember that this use of screening-level concentrations in the risk-based screening method identifies priority chemicals, but it does not measure the actual risk that these chemicals present to the community. Risk is determined by the toxicity of a chemical and by the amount of exposure to the chemical. To save on resources, risk-based screening uses estimated air concentrations as a surrogate for exposure. It does not attempt to measure the actual amount of exposure that members of the community have to the priority chemicals. Estimating exposure is a key step in determining potential health risk. People move around from one location to another, outside to inside, so exposure isn't the same as the concentration at a static site. People also breathe at different rates depending on their activity levels, so the amount of air they take in varies. For these reasons, the average concentration of a pollutant that people breathe (i.e., exposure concentration) may be significantly higher or lower than the concentration at a fixed location (i.e., the estimated ambient concentration).

Keep in mind that the estimated air concentration that the Partnership uses as a surrogate for the exposure concentration is just one of the factors used to calculate screening-level concentrations. The conservative nature of other factors used to calculate the screening concentration, such as the duration of exposure and the hazard value for the chemical, is likely to be protective of sensitive populations at the level of risk chosen by the Partnership. If the Partnership is concerned about using screening calculations that are protective of sensitive populations, it has the option of choosing a more conservative risk level to ensure protection. A partnership can also use a less conservative risk screening level to identify priorities, evaluate the priority chemicals identified by the screen to find feasible risk reduction actions, begin work on risk reduction actions, and then, as risk reduction work proceeds, reevaluate the screening concentrations to see if they need further adjustment to protect sensitive populations.

If a community, in addition to identifying priority chemicals, wants information on the amount of risk these chemicals present to the community, they will need to complete a risk assessment. A discussion of this option and a reference to guidance for conducting a risk assessment can be found in the section on choosing a work plan in Chapter 2. There will also be further discussion of this issue in the Secondary and Final Screen chapters of this Manual. At these steps of the screening process, the more accurate methods used to estimate concentrations provide better information on ambient concentrations that could be used to estimate exposure and risk.

How will the Partnership calculate the screening-level concentrations for each of the chemicals in community air?

Because the work to develop these screening-level concentrations is technical, the Partnership committee will need to appoint a technical team to calculate the screening-level concentrations that it will use. This technical team will need to find and use information on toxicity for each chemical found in community air. Once the technical team has assembled the necessary toxicity information for each of the chemicals in community air, the team will calculate the risk screening concentration by combining the Partnership's risk screening level with the toxicity value found for each chemical.

Here is an example of the calculation for arsenic, a cancer causing chemical, and acrylic acid, a non-cancer-causing chemical.

EXAMPLE 1: Arsenic compounds as cancer screening-level concentration (SLC) example

$$\text{SLC (cancer) } \mu g/m^3 = \frac{1.0E\text{-}06}{4.3E\text{-}03 \text{ per } \mu g/m^3 \text{ (from IRIS)}} = 2.3E\text{-}04 \ \mu g/m^3$$

EXAMPLE 2: Acrylic acid as non-cancer screening-level concentration (SLC) example

$$\text{SLC (non-cancer) } \mu g/m^3 = 1 * 1.0E\text{-}03 \text{ mg/m}^3 \text{ (from IRIS)} * 1000 \ \mu g/mg = 1.0 \ \mu g/m^3$$

The Partnership's technical team will be able to explain these examples in detail, including the technical notation used to express the numbers. The toxicity values used in these calculations come from a database developed by EPA's Office of Air Quality Planning and Standards (OAQPS) accessible at http://www.epa.gov/ttn/atw/toxsource/summary.html.

For cancer-causing chemicals, the unit risk estimates used to calculate the screening-level concentrations are considered plausible upper-bound estimates of an individual's probability of contracting cancer over a lifetime of exposure, meaning that they represent a plausible upper limit to the true value. The true risk is likely to be less but could be greater. As a result, the risk that would result from exposure to the screening-level concentration is also an upper-bound estimate of risk.

The true level of risk that would result from exposure to the screening-level concentration is likely to be lower than the community risk screening level, but it could be greater. The Partnership can rely on its technical team to explain the use of upper-bound estimates of risk.

The method for calculating screening-level concentrations described above differs from the method used in Baltimore. In Baltimore, the Partnership used risk based concentrations that are available in EPA Regional databases set up for the Superfund program. These risk based concentrations were calculated using a different approach for cancer-based values. The method suggested above is recommended to take advantage of current Agency methodology and a database of toxicity information developed by EPA's Office of Air Quality, Planning, and Standards (OAQPS) that was not available for the Baltimore project.

Could the Partnership use a proportion of risk to screen chemicals as an alternative or supplement to the use of screening-level concentrations?

There is an alternate method for conducting a risk-based screening that uses a proportion of total risk to screen chemicals instead of screening level concentrations. The Partnership in Baltimore, after considerable discussion, decided that screening-level concentrations would be easier to explain to the community and would be less likely to be mistaken for estimates of true risk, so the proportion-of-risk screening approach was not used in Baltimore. Consequently, only a brief description of how to use proportion of risk as a screen is provided here. If you decide to use this approach, you will have to rely on the technical members of your Partnership to provide detailed guidance. If your Partnership adopts this approach, please summarize and share your experiences so that other communities can learn from them and more detailed guidance can be developed.

To use proportion of cumulative risk to screen chemicals, a partnership would first estimate exposure concentrations or might agree to use ambient air concentration estimates as surrogates for exposure. Then, for each chemical with cancer-based toxicity values, the Partnership would multiply the estimated ambient concentration by the chemical's unit risk estimate to get the risk value. Since the estimated ambient air concentration is a surrogate measure of exposure, the risk value produced by this method would not be an estimate of true risk in the community. But this risk value could be used to develop a relative ranking of chemicals by risk.

To ensure the quality of its analysis, the Partnership will want to make sure that its technical team finds and uses the best available data

As an alternative to using risk screening concentrations to identify the chemicals needing further analysis, the Partnership could rank chemicals by their relative risk and pick the chemicals with the highest relative risk for further analysis in the Secondary Screen and as priorities to be evaluated for possible community action in the Final Screen. The Partnership could also choose a proportion of the total of all risk values calculated for all the chemicals to screen chemicals. The Partnership could calculate the percentage that each chemical contributes to the total risk value associated with all chemicals in this assessment and rank the chemicals by percentage of risk they contribute to the total. The Partnership could choose a percentage of risk for use during the screen and screen chemicals using this percentage. For example, if a partnership chose 95% as its screening percentage, it would start with the chemical contributing the greatest relative risk and go down the ranked list of chemicals until 95% of the total of the risk values is accounted for. These chemicals would be kept for further review and the chemicals contributing the remaining 5% of the total of the risk values would be dropped. If a partnership uses this method to screen chemicals, it would have an idea of the proportion of total air toxics risk that the community is addressing once it has completed its screen and identified its priority chemicals and sources. The ability to demonstrate that a high proportion of the risk will be addressed by working on the priorities identified by the screen may increase the community's confidence in the screening method.

How will the Partnership provide guidance to its technical team in choosing the sources of chemical toxicity information?

Since toxicity information is available in many databases, and since the information on toxicity in these databases may differ, the technical team will need to select the toxicity information that will be used to calculate screening-level concentrations. To ensure the quality of its analysis, the Partnership will want to make sure that its technical team finds and uses the best available information. The easiest way to identify the best available information may be to decide on a hierarchy for the available toxicity databases that ranks them, according to the quality of the data they contain, from the database with the highest quality of information to the database with the lowest. The toxicity information from the highest ranking database with information on a chemical can then be used for calculating screening-level concentrations.

The quality of the information in databases depends on factors such as the level of peer review and the currency of the data. The level of peer review is an indication of the level of review by independent scientists that the data included in a database have undergone. Peer review levels can range from complete review of all data by independent scientists, to review by scientists within the same organization publishing the data, to no peer review. The Partnership's confidence in the data contained in a database will vary depending on the amount and kind of peer review a database has undergone. The Partnership's technical team will probably not be able to find the information they need to calculate screening-level concentrations for all of the chemicals in community air if they use only the databases with the highest level of peer review. The information available on many chemicals has not undergone the highest level of peer review.

Currency of the data refers to the time when the information on a chemical was developed and reviewed. New testing and new information on chemicals is being developed all the time, so the information on chemicals in databases needs to be updated regularly.

To assist with efforts to understand risks associated with air toxics, EPA's Office of Air Quality Planning and Standards (OAQPS) has developed a method of ranking databases and has combined available toxicity information for hazardous air pollutants in one database, available on the web at http://www.epa.gov/ttn/atw/toxsource/summary.html.

To create this database, EPA's OAQPS established a hierarchy of sources of dose-response or toxicity assessments based on the quality of the data they contain and then used this hierarchy to select the toxicity value to include in their database. Links are provided in this web site to tabulated dose-response assessments that OAQPS uses for risk assessments of hazardous air pollutants. Tables are provided with values for long-term (chronic) inhalation and oral exposures and for short-term (acute) inhalation exposures. The tables compile assessments from various sources for many of the 188 substances listed as hazardous air pollutants ("air toxics") under the Clean Air Act Amendments of 1990. Each assessment in these tables is best visualized as an estimate within a range of possible values, surrounded by uncertainty and variability. This range of possible values may change as better data become available.

We recommend using this database as the primary source for the toxicity information needed to calculate the Partnership's screening-level concentrations. Using this database not only will make the job of the technical team easier and ensure the quality of the data used, it will also make it possible for the Partnership to compare the results of its risk-based screening to other communities using the same database. But keep in mind that the quality of information in this toxicity database, like all databases, varies from chemical to chemical. For chemicals identified as priorities, the Partnership's technical team may want to examine the quality of the toxicity value and summarize this information to help inform any decisions the Partnership might make that would be based on this information.

In general, it is important to remember that the quality of the data contained in toxicity databases can vary from chemical to chemical depending on the level of peer review and currency of the information. In addition, the OAQPS database and other databases, for example, the Agency for Toxic Substance and Disease Registry's **Minimal Risk Levels (MRLs) for Hazardous Substances** and the California Environmental Protection Agency's **Toxicity Criteria Database**, all contain some degree of uncertainty because of the lack of precise toxicological information on the people that might be most sensitive (e.g., infants, elderly, and nutritionally or immunologically compromised) to the effects of hazardous substances. To address these uncertainties, toxicity values are calculated using a conservative (i.e., protective) approach.

Chemicals in community air that are not on the Clean Air Act list of hazardous air pollutants may not be included in the OAQPS toxicity database (e.g., aldicarb, allylamine, benzamide, bifenthrin, bromine, carbofuran, creosote, diazinon, dichlorobromomethane). For these chemicals, the Partnership's technical team will need to search other available toxicity databases to find toxicity information. To ensure that the technical team uses the best available information, the Partnership will need to establish a ranking of the available databases and instruct its team to use the information from the highest ranked database. If there are chemicals in community air that are not included in the OAQPS database, the Partnership's technical team can present the Partnership with a plan for finding the highest quality data for discussion and approval. To be consistent, the Partnership may want to adopt the hierarchy of data sources used by OAQPS to develop its database for the 188 Clean Air Act hazardous air pollutants.

As the Partnership's technical team develops screening-level concentrations and enters them into the Partnership's database, it would be helpful if the team also indicated the source of the information used so that the Partnership would be able to easily consider the strengths and weaknesses of the assessment that underlies each of its screening-level concentrations. This information will be especially important at the end of the screening process if a chemical is identified as a priority and the Partnership needs to consider possible actions based on this information.

Why is it important for the Partnership to agree on and set clear standards for choosing the sources of data that its technical team will use?

The goals of the Partnership's screening exercise are to reach agreement in the community on the chemicals that need attention and to mobilize the community to take effective action to reduce the risks from these chemicals. If everyone in the Partnership agrees at the start of the screening exercise that the decisions made in setting screening-level concentrations are reasonable and valid, then the Partnership will have a sound basis for taking action on the chemicals identified as priorities. Taking time to thoroughly discuss and reach agreement on all the decisions that will be made in the screening process will help to avoid the problem of second-guessing the work of the Partnership once the results are in and will help to avoid any questions on the screening methods that could delay the community's efforts to improve air quality.

Taking the time to thoroughly discuss and reach agreement on all the decisions that will be made in the screening process will help to avoid the problem of second-guessing the work of the Partnership once the results are in

What are all the decisions and steps required to develop the screening-level concentrations that the Partnership will use?

Here is a summary of the decisions and steps that the Partnership and its technical team will need to complete to develop the screening-level concentrations.

Summary of Steps to Develop Screening-Level Concentrations

Step 1: Partnership appoints a technical team to calculate risk screening concentrations following QA/QC procedures

Step 2: Partnership provides the education that its members will need to participate fully in the decisions that will need to be made to ensure that the work to set screening-level concentrations meets community goals and values

Step 3: Partnership decides on the risk screening level that it will use to trigger possible action

Step 4: Technical team develops a plan for accessing toxicity information and calculating screening-level concentrations to present to the Partnership for discussion and approval

Step 5: Partnership approves plan for accessing toxicity information and calculating screening-level concentrations

Step 6: Technical team develops risk screening concentrations for the Partnership and stores them in the Partnership database

Step 3: Comparing Community Air Screening Concentrations to the Screening-Level Concentrations

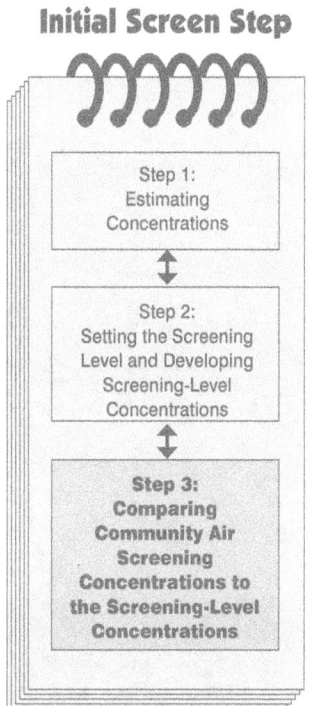

Initial Screen Step

Step 1:
Estimating
Concentrations

Step 2:
Setting the Screening
Level and Developing
Screening-Level
Concentrations

Step 3:
Comparing
Community Air
Screening
Concentrations to
the Screening-Level
Concentrations

What does the Partnership do once it has screening-level concentrations for the chemicals in community air?

The final step of the Initial Screen is to compare the estimated concentration for each chemical to its corresponding screening-level concentration. The Partnership uses this comparison step to identify all the chemicals that have estimated concentrations that exceed their corresponding screening-level concentrations. The Partnership identifies these chemicals as candidates for further analysis. Chemicals with concentrations equal to or less than their screening levels are eliminated from further review in this screening exercise.

Why should the chemicals kept for further review by the Initial Screen not be described as community priorities?

The conservative assumptions that are built into the Initial Screen, described in the introduction to this chapter, mean that the concentrations estimated for the Initial Screen are likely to be overestimated. As a result, the actual concentrations in community air of the chemicals identified by the Initial Screen may not be above the Partnership's risk screening concentrations. Only the more accurate information and methods that will be used to estimate concentrations in the Secondary and Final Screening steps will be sufficient to determine if a chemical will be identified as a community priority. Until these more accurate estimates are developed, the Partnership will not have enough information to identify its priorities. The primary purpose of the Initial Screen is to find and eliminate the chemicals that the Partnership can confidently assume are below the community screening concentrations. This allows the more resource-intensive effort needed to produce more accurate estimates in the Secondary and Final Screens to focus on the remaining chemicals that may be above the screening levels. Bottom line: The chemicals identified during the Initial Screen must be further evaluated during the Secondary and Final Screening steps before they can become candidates for possible community action.

Overview

Secondary Screen

Organize
Collect
Analyze
Mobilize

What is the purpose of the Secondary Screen?

The Secondary Screen is designed to take the chemicals left after the Initial Screen and, using a more sophisticated method to estimate concentrations, once again sort through them to find those that can be set aside and those that will require further review. This is the second step in the risk-based screening process designed to identify the chemicals that will become community priorities.

The Secondary Screen will also, for the first time, review the chemicals released by mobile sources and by stationary point sources that are closer than 15 meters to the nearest community exposure. All the chemicals released by these sources were passed through the Initial Screen and kept in the screening process for review in the Secondary Screen.

How does the Secondary Screen work?

The screening steps outlined in this Manual were designed to find an effective way to sort through a large number of chemicals and sources with the minimum expenditure of resources. The Initial Screen uses a simplified method to estimate concentrations so that all the chemicals in community air can be reviewed efficiently and the chemicals with concentrations obviously below community screening levels can be eliminated from the screening. With a smaller set of chemicals left after the Initial Screen, it is now possible for the Partnership to use a more resource-intensive method to get a more accurate estimate of the concentrations in community air. The more accurate concentrations estimated in the Secondary Screen are, once again, compared with the screening-level concentrations that were set for the Initial Screen. Chemicals whose new estimated concentrations are still above the community screening levels will need further review and will be kept for the Final Screening. They will remain as possible candidates for the community's list of priority chemicals. The chemicals whose more accurately estimated concentrations are now at or below the community screening concentrations can be set aside. They will not need further review in the Final Screen.

What does the Partnership do in the Secondary Screen to get more accurate estimates of community air concentrations?

For the Secondary Screen, instead of using the conservative assumptions and the simplified SCREEN3 look-up table, the Partnership will use the ISCST air dispersion model to provide more accurate estimations of concentrations in community air.

What is the ISCST air dispersion model and what does it do?

The ISCST air dispersion model is a widely used model developed by EPA to estimate the air concentrations that result from chemical releases. ISCST can be downloaded from an EPA web site and run on a standard desktop computer. Please see the general explanation of air dispersion modeling in the Initial Screen section of the Overview. In the future, communities will have the option of accessing and using this model via the Internet using the Internet Geographic Exposure Modeling System (IGEMS), a program under development by EPA. Information on IGEMS can be accessed at http://www.epa.gov/oppt/exposure/.

To use the ISCST air dispersion model, the Partnership will enter information on the location and characteristics of the sources, the releases, and the local meteorological conditions from the nearest meteorological station into the model. Using this information, the model will predict the dispersion of chemicals from the sources (predict the plumes) and estimate the concentrations that will result from the releases at selected points in the Partnership area. If a chemical is released by more than one source, the ISCST air dispersion model will calculate the concentration that will result from all the sources combined together at the selected points in the Partnership area. The Partnership will decide on the locations in the community that are of special concern, such as schools, nursing homes, and residential areas, and the model will estimate the concentrations of chemicals from all sources at these locations. The model can also be used to identify any locations in the community that have estimated concentrations that are above the community screening level. The model will estimate the concentrations in community air for each of the chemicals identified by the Initial Screen.

What background information will the Partnership need to provide its members to ensure that they have the ability to oversee the technical work and communicate the results to the community?

At this point, it will be important for the full Partnership to take time to make sure that all of its members have a good understanding of the value and limits of air dispersion modeling in general, and the ISCST model, in particular. This background will be essential both to ensure that the Partnership has the ability to oversee the technical work of the screening and to communicate that work to the broader community. Members of the Partnership with experience in explaining technical information can take the lead in providing this background. If necessary, the Partnership could request help from outside the Partnership to provide this education. As a part of this education, a demonstration of the model and its use may be especially helpful to the Partnership. As discussed below, the Partnership will need to appoint a technical team to carry out the air dispersion modeling for the Partnership. Although the task of this team will be to complete the technical modeling work for the Partnership, the members of this technical team could assist or take the lead in the educational work as well. EPA is currently developing educational materials for use in the classroom to explain and demonstrate air dispersion modeling. These materials will be available to help communities with educational efforts in the future.

It will be important for the full Partnership to take time to make sure that all of its members have a good understanding of the value and limits of air dispersion modeling

Why does the ISCST air dispersion model provide more accurate estimates of concentrations than the Initial Screen?

The method used to estimate concentrations in the Initial Screen is designed to overestimate concentrations. It does this by assuming that the meteorological conditions that result in the greatest concentrations will remain the same for every day of the year and by assuming that the sources are all in the same exact location so that the highest concentrations resulting from each of the sources can be added together. The ISCST air dispersion model, in contrast, uses the actual meteorological conditions as measured at the weather station closest to the Partnership area and the actual locations of sources to estimate concentrations. With this more realistic information, the model can estimate the dispersion of releases from each source and provide a more realistic estimation of the combined concentrations that occur if the plumes from the sources overlap.

With the more realistic input, the concentrations estimated using the ISCST Air Dispersion Model should be lower than the estimates made in the Initial Screen and closer to the actual concentrations in community air. But, keep in mind, for stationary sources these new estimates of concentrations are still designed to be conservative, i.e., overestimates, because in the Secondary Screen the Partnership will still be using readily available release data based on maximum permitted release amounts as input to the model. For the next, and final, screening step, with the number of chemicals left in the process further reduced, the Partnership will have the resources to contact each of the remaining stationary sources to obtain the most accurate release data. The Final Screening step will be explained in detail in the next section of the Manual.

If the release data available to a community Partnership only has estimates of actual release amounts and not maximum permitted release amounts for stationary sources, the Partnership will need to increase these release estimates, possibly multiplying the release amount by a factor agreed on by the Partnership, to ensure that the Secondary Screen is conservative. Release estimates taken from the TRI database, for example, are estimates of actual releases, not maximum permitted releases, so the Partnership's technical team will need to increase these amounts to ensure that the Secondary Screen is conservative.

Are there any special characteristics of a community that need to be accounted for in air dispersion modeling?

The ISCST model has many features and options that allow the model to estimate dispersion in a variety of special situations. This Manual, following the principle of effectively identifying community priorities with a minimum expenditure of community resources, does not provide guidance for using all of the detailed modeling capabilities of the ISCST model. The Manual recommends the use of a simplified version of ISCST modeling to estimate concentrations. In some communities, this simplified use of the ISCST model may not be adequate to estimate concentrations.

There are two situations, in particular, that may call for your Partnership to use more sophisticated modeling to effectively estimate concentrations: **complex terrain**, such as a community located in a valley surrounded by high hills or on hillsides higher than the sources of emissions, and the presence of **high buildings** close to a source or sources. Unusual terrain features can cause changes in wind patterns that affect dispersion as well as shift the position of the receptor closer to the higher concentrations in the plume.

High buildings close to sources can cause an effect called "building downwash" that results in higher concentrations close to the building. References to guidance for identifying when and how to use ISCST to account for terrain and buildings are provided for your technical team in the Technical Guidance section of the Manual. If your community may fit one of these special circumstances, you will need to work with your technical team to be sure that you are using adequate air dispersion modeling. In some unusual circumstances, you may even need to consider use of a different air dispersion model to effectively estimate concentrations. The Partnership, with the help of its technical team, should discuss this issue and agree on the plan for modeling to make sure that everyone has confidence in the results of modeling.

The guidance provided in this Manual does account for differences in the dispersion of releases that are in the form of particles rather than gases. Particles tend to settle more rapidly than gases, so their dispersion pattern can be significantly different than the pattern for a gas. Releases of metals and many releases resulting from combustion, including releases from the burning of diesel fuel, are more effectively modeled as particulates. Since particulate releases will be part of the inventory for most, if not all, communities, the Manual includes guidance for your technical team to identify particulate releases and use the capability of the ISCST to model their dispersion.

What will the community need to keep in mind to understand the limits and the value of using the ISCST air dispersion model as a part of the screening process?

The ISCST model, like all computer models, must simplify the extremely complex process of the dispersion of chemicals in community air to make a tool that is practical to use. The modeling approach used in this Manual, for example, does not take into account factors that affect air concentrations, such as the daily variations in release that occur from some sources or the reactions that some chemicals undergo once they are released in the air. In addition, the Secondary Screen, like the Initial Screen, still uses overestimates of releases. Perhaps most important, the modeling results will reflect any uncertainty inherent in the information that the Partnership has collected in its inventory database and used as input to the model, such as the stack characteristics and locations of emission sources.

As a result, the Partnership should view the concentrations estimated by the ISCST model as only approximate estimates of the actual concentrations in community air. Studies that have compared ISCST modeling estimates to measured concentrations have shown that modeled estimates, depending on the situation, can differ from measured concentrations by a factor of 2. (Please see the Technical Guidance section, Chapter 11, for some discussion of and references for these studies.) Despite its inherent uncertainties and limits, air dispersion modeling does provide communities with a valuable tool to help set priorities. Modeling allows a community to estimate concentrations with **enough accuracy to effectively identify the chemicals that should be the focus for possible community action.** The ability to identify and agree on priorities can be crucial to the effective use of community resources to improve air quality. Using modeling to reach agreement on community priorities, like all the methods and information used in the screening process, depends on everyone in the Partnership understanding the uncertainties involved in the screening process and agreeing in advance that the screening method is reasonable and sufficient to identify community priorities. This agreement depends, in part, on the recognition that the alternatives to modeling, such as monitoring, would be significantly more expensive and time consuming and would very likely result, given the level of accuracy of modeling, in identifying the same priorities. Please refer to the discussion of alternate technical approaches for analyzing air quality in Chapter 2.

Understanding the level of uncertainty inherent in modeling will also help the Partnership to interpret and explain the results of the modeling. Modeling uncertainty will be especially important to remember when reviewing the concentration numbers produced as the output of the model. Because the model results will be expressed in the form of numbers with two, three, or more decimal places, such as $0.156\ \mu g/m^3$, they can look very accurate. They are not that accurate. In fact, in communicating the results of the modeling to the community, the Partnership may want to find a way of expressing the results that better reflects the uncertainties inherent in the modeling.

And an understanding of the uncertainties inherent in modeled estimates will encourage the Partnership to consider using its own judgement in deciding which chemicals to move to the final step of the screening process. The Partnership may decide to keep a chemical in the process despite the fact that its estimated concentration is below the screening level. For example, if the estimate of a chemical's concentration is close to, but below, the screening level and there are few chemicals above the screening level, the Partnership could decide, realizing the uncertainty in the modeled estimate, to keep the chemical for further review in the Final Screening step. On the other hand, if there are many chemicals with concentrations significantly above their screening value, the Partnership may decide to drop a chemical with a concentration that is above, but close to, its screening value. Understanding the nature of modeling, and the uncertainties included in the estimates it produces, will allow the Partnership to make the kinds of informed judgments that will increase the effectiveness of the overall screening process. Of course, these judgments need to be made with the participation and agreement of the full Partnership committee so that everyone will be in agreement on any actions that may result from the screening effort.

This may also be a good point to remember that the air dispersion model estimates air concentrations, not the exposure of members of the community to these concentrations. Exposure is a measurement of how much an individual or a community breathes the outdoor air concentration. Measuring this exposure would require information on factors such as the amount of time spent outdoors, the years spent in the community, rates of breathing, etc. Since the methodology described in this Manual does not include the collection of this exposure information, the community cannot use the estimated concentrations to estimate the risk that these chemicals pose to the community. See the fuller discussion of this issue in the section on screening-level concentrations in Chapter 5.

How will the Partnership find the technical skills and resources that will be needed to use the ISCST air dispersion model?

The Partnership technical team assigned to collect the information needed for the project and the team assigned to estimate concentrations will work together to complete the Secondary Screen. The team estimating concentrations will need members with skills in using air dispersion modeling. Partnership members from colleges and universities, industry and government technical staff, or retired community professionals are likely to have the skills needed to use the ISCST model. This technical team will need access to a computer and to the ISCST model. The technical team members will need time to set up and run the model with the information that the Partnership has collected. If the members of the technical team do not have the time or the skills to run the air dispersion model themselves, the Partnership may need to find funding to hire a contractor to run the model for the community. In this case, a technical team from the Partnership will probably still be needed to oversee the work of the contractor.

How will the ISCST model handle the different types of sources and what new information will be needed to complete this modeling?

The following section provides a general overview of how the ISCST air dispersion model will be used to estimate the air concentrations for the different types of sources in the community. Understanding these approaches will help to explain the different kinds of information that will be needed to complete the Secondary Screen.

For the stationary point sources: The location of each of these sources will be entered into the ISCST model, and the dispersion of the releases from these facilities will be modeled individually from their locations in the community. The concentrations resulting from the releases of each facility will be estimated throughout the Partnership area.

Information required: For the stationary point sources, the model will need information on the location of the source, the chemicals released, the amount of each chemical released, the release characteristics needed for modeling such as stack height and the velocity of the release as it leaves the stack, and local meteorological information.

For the stationary area sources: Since these sources are too numerous or dispersed to model individually, air concentrations will be estimated by apportioning the total county-wide emissions to each census tract in the community and by assuming that the apportioned emissions are released from five imaginary points (pseudo-points) located in each census tract. The ISCST air dispersion model will estimate the concentrations resulting from the releases from these pseudo-point sources.

Take, for example, the releases from the heating of all the homes in the Partnership area. Since there are too many homes to locate and model individually, the Partnership's technical team will use the estimated emissions for the entire county and apportion them appropriately to each of the census tracts in the Partnership area. In this case, the surrogate or factor used to apportion emissions from home heating is population. The Partnership will compare the population in a census tract to the total population in the county and assign an equal portion of the total county releases to that census tract. Once the releases have been apportioned to each census tract, the concentrations that would result from these releases will be estimated by assuming that the releases come from five imaginary points spread evenly throughout the census tract. In other words, the concentrations will be estimated by modeling from five imaginary points instead of the much larger number of homes actually located in the census tract. Modeling the concentrations from the five imaginary points in each census tract will provide an estimate of the concentrations resulting from the releases from household heating. This basic modeling approach will be used to estimate the concentrations from all the types of sources too numerous to model individually.

Information required: To estimate the concentrations resulting from the stationary area sources, the Partnership will need census tract geographic information and census tract centroids, county-wide estimated releases for each source type, chemicals released, information needed to apportion releases to each census tract, and local meteorological information.

For the on-road and non-road mobile sources: For both types of mobile sources, the technical team will follow the same procedure used for stationary area sources described above. A proportion of the county-wide total emissions for the on-road and non-road mobile sources will be apportioned to each census tract, and concentrations will be estimated by assuming that the apportioned releases come from five imaginary point sources and by modeling the concentrations that would result from these five imaginary points.

Information required: The information needed to complete this task includes county-wide release data, census tract geographic information and census tract centroids, a method for apportioning county-wide release estimates to each census tract, and local meteorological information.

Detailed guidance to help the team use the ISCST model to estimate concentrations for each type of source can be found in the Technical Guidance section of the Manual. The Inventory Team and the team estimating concentrations can work together to collect the new information that will be needed. Guidance to help the teams find this data can also be found in the Technical Guidance section.

The information needed to run the ISCST model in the Secondary Screen is summarized in Table 6-1.

Table 6-1. Summary of Information Needed to Run the ISCST Model in the Secondary Screen

TYPE OF SOURCE	INFORMATION NEEDED
STATIONARY SOURCES	
Stationary Point Sources: All large and small commercial, institutional, and industrial sources (sources will be modeled individually)	• Chemicals released • Release amounts • Location of source • *Modeling release parameters* • *Local meteorological information*
Stationary Area Sources: Residential, building, and other miscellaneous sources (sources combined for modeling)	• Chemicals released • Total county emissions • *Census tract geographic information and centroids* • *Information needed to apportion county emissions* • *Local meteorological information*
MOBILE SOURCES	
On-Road	• Chemicals released • Total county emissions • *Census tract geographic information and centroids* • *Information needed to apportion county emissions* • *Local meteorological information*
Non-Road	• Chemicals released • Total county emissions • *Census tract geographic information and centroids* • *Information needed to apportion county emissions* • *Local meteorological information*

Note: The new information needed to complete the Secondary Screen is highlighted in italics.

What are all the steps the Partnership will need to take to complete the Secondary Screen?

The following is a list of all the steps the Partnership will need to complete during the Secondary Screen.

Summary of Steps to Complete the Secondary Screen

Step 1: Appoint a technical team with the skills needed to use the ISCST model to estimate concentrations.

Step 2: Provide education on air dispersion modeling to the full Partnership committee so that all members understand the role and the limits of the ISCST model that will be used in the Secondary and Final Screening steps.

Step 3: Set the quality control system to check the analysis to ensure that it follows Partnership guidelines and that it is accurate and complete.

Step 4: Complete the collection of all the information needed for running the ISCST model.

Step 5: Choose the locations in the community where the ISCST air dispersion model will estimate air concentrations.

Step 6: Acquire or get access to the use of the ISCST model and a computer to run the model.

Step 7: Set up and use the ISCST air dispersion model to estimate concentrations.

Step 8: Compare the new concentrations estimated for the locations identified by the Partnership to the community screening-level concentrations and identify those above the screening concentrations.

Step 9: Communicate results to the broader community.

Are there any key decisions that the full Partnership needs to make to ensure that the Secondary Screen meets community needs?

At this point, most of the decisions needed to ensure that the screening process meets community needs, such as deciding on the risk screening level and setting standards for selecting data, have already been made. Work during the Secondary Screen will continue to follow the guidelines set by the partnership in the previous step. For this step, the Partnership will need to identify the locations of special concern in the community and instruct its technical team to use the air dispersion model to estimate the concentrations at these locations.

What should the Partnership consider in choosing the community locations where the air concentrations will be estimated?

Working with the broader community, the Partnership may want to identify locations in the community where information on air concentrations will be especially important. These can include locations where community members who may be especially sensitive to air toxics might be exposed. Schools, playgrounds, sports fields, and day care facilities could be identified for children, and nursing homes and assisted-living facilities could be identified for older community members. Residential areas where community members are concentrated and residences close to sources of toxics, either stationary or mobile sources, may also be a concern. Identifying locations with special concerns and providing information on these locations to the community will help to ensure that community questions are answered directly. In addition to estimating concentrations at specific locations, keep in mind that the ISCST model will provide information that will allow the Partnership to identify any area in the community with a concentration above the community screening level.

The Partnership will need to draw on all of its members to identify locations in the community that are of special concern. Community members, local schools, and local public health practitioners should be especially helpful in this discussion.

Please note that the modeling methods used in this Manual result in some significant limitations in the ability to estimate the aggregate concentrations from all sources at specific community locations. It will be important for the Partnership to understand and explain these limits to the broader community, especially following the Final Screening step, when community priorities have been identified. In particular, the pseudo-point method used to estimate concentrations for stationary area sources (e.g., home heating) and for mobile sources tends to spread the releases evenly across each census tract. For mobile sources especially, this may not adequately reflect concentrations at points close to roads or intersections with heavy traffic. In other words, the concentrations close to mobile sources at some community locations may be underestimated by the pseudo-point method. This means that for those community locations close to mobile sources, the aggregate concentration that results from combining the mobile source estimates with the stationary point and area source estimates may also be underestimated. This possibility and the limits in the ability of the pseudo-point method to identify mobile source hotspots need to be kept in mind and explained to the broader community. For the purposes of the Secondary Screen, if the pseudo-point method used to estimate mobile source chemicals results in aggregate concentrations that are below screening-level concentrations, the Partnership may need to consider moving these chemicals to the Final Screen where a more accurate modeling method will be an option. In most cases, the pseudo-point method of estimating mobile sources, despite the fact that it underestimates concentrations in locations close to mobile sources, will result in concentrations above community screening levels. Please see a further discussion of mobile sources in Chapter 7.

How long will it take to complete the air dispersion modeling using the ISCST model?

The time needed to complete the air dispersion modeling will vary depending on the number of chemicals and sources that need to be modeled and the resources that are available to set up and run the model. If all the information is collected and ready to use, the Partnership's technical team will spend most of its time setting up the model with all the information on locations and releases of sources, locations of concern, and census tract boundaries. In Baltimore, it took about 40 hours of work to set up the model for the study neighborhoods and to input all the information for the 29 chemicals and 36 sources that were modeled in the Secondary Screen for stationary commercial and industrial sources only. Setting up the model to add household and mobile sources in Baltimore would have required additional time, perhaps as much as 40 more hours. Once all the information was entered into the model, running the model to estimate the concentrations took only several hours. This too may vary depending on the speed of the computer used.

It is good to know that the investment of time and resources required to set up and enter the information on sources and their locations into the model will need to be done only once. All the work done to set up the model to reflect community details can be saved. Future community modeling will require only updating the model with any changes that have occurred since the last modeling.

How could the Partnership carry out the technical work of the Secondary Screen in a way that helps to build the long-term capacity of the community to understand and improve air quality?

Training teachers and students from local schools or training residents would help to ensure that the capacity needed for future updates of the modeling will be available in the community

The partnership and its technical team may consider developing the capacity of a community group and/or community individuals to do the air dispersion modeling needed for the Secondary Screen. Instead of doing the modeling themselves, technical team members could train and work with local science teachers and their students, a local youth organization, or a local group of retired professionals to carry out the modeling. Training teachers and students from local schools or training local residents would help to ensure that the capacity needed for future updates of the modeling will be available in the community. Providing training for a community group would also help the Partnership develop the skills needed to communicate the modeling work to the broader community. Providing schools with the opportunity to work with the Partnership and use air dispersion modeling to help answer community questions about local air quality could create an ideal setting for the teaching of science in neighborhood schools.

In addition to the skills needed for air dispersion modeling, what other skills will the Partnership need to complete the Secondary Screen?

As in the Initial Screen, the Partnership will need to draw on the skills and resources of most of its members to complete the Secondary Screening step. In addition to the technical skills needed to run the ISCST air dispersion model, discussed above, the Partnership will need the following.

Science education skills: The full Partnership committee will need to understand how the ISCST air dispersion model works and how it will be used in the Secondary Screen. Committee members with skills in communicating technical information will be needed to make sure that all the Partnership members have the information and understanding they will need to participate fully in oversight of the technical team's work and in the use and interpretation of the results of the modeling. Science teachers from local schools, colleges, and universities, and any technical members of the committee with experience in communicating technical information, could provide these skills.

Communication skills: Explaining the Secondary Screen and its results to the broader community will be an important part of the Partnership's work during and at the completion of the Secondary Screen. (See discussion of communication below.) The Partnership committee will need individuals with the ability to communicate the work of the Secondary Screen to the community. Community members, teachers, journalists, or possibly a team that combines community members with the professional writers could provide these skills.

Does the Secondary Screen identify the chemicals and sources that will become the community's priorities?

Since the air dispersion model in the Secondary Screen still uses maximum permitted releases, the Secondary Screen is designed, like the Initial Screen, to overestimate concentrations. While the Secondary Screen is more accurate and less conservative than the Initial Screen, it is still intended, except in the case of mobile and stationary area sources, to overestimate concentrations.

As a result, the actual concentrations of the chemicals identified by the Secondary Screen are likely to be lower than the estimated concentrations. This means that the Secondary Screen is not designed to identify the community priority chemicals. Like the Initial Screen, the Secondary Screen only identifies chemicals and sources that deserve further consideration. In the Final Screen, the Partnership will work to get the best information available to estimate concentrations. This next and final screening step will identify the chemicals and sources that will be the community's priorities. Both the Initial and Secondary Screens are designed to provide a resource-effective method to eliminate chemicals that are very likely to be at or below community screening levels so that the resource-intensive effort of visiting stationary point sources to collect detailed information on releases can be focused on only those chemicals that need this detailed review. The Final Screen, described in the next section provides that detailed review.

What can the Partnership do to communicate the results of the Secondary Screen to the broader community?

Explaining the work on a regular basis throughout the project will ensure that the Partnership develops the ability to communicate effectively with the community

As discussed in Chapter 2, it is important to communicate with and involve the broader community at every stage of the screening process. The Partnership could take the occasion of the completion of this phase of its work as an opportunity to develop a press release and send speakers to community groups to explain the work. Of course, it will be important to explain that the chemicals identified at this stage of the work are only candidates for further review and that they may not have actual concentrations above the screening values. The Partnership can also take this opportunity to explain that the chemicals identified in the Secondary Screen will be reviewed in the Final Screen with the best available information, and the results of this review will be used to identify the community's priority chemicals.

Developing educational materials and explaining the work on a regular basis throughout the project will ensure that the Partnership develops the ability to communicate effectively with the community and that the community is aware of and supports the work of the Partnership. It will also be important to take this opportunity to make sure that all the sources of the chemicals identified in the Secondary Screen have been contacted and encouraged to join the Partnership.

Overview **7**

Final Screen

What is the purpose of the Final Screen?

The Final Screen is designed to take the chemicals still in the process after the Secondary Screen and, by collecting and using the best information on sources that can be collected, develop new and more accurate estimates of concentrations in community air. With these newly estimated concentrations, the Final Screen will be used to identify the chemicals and the sources that will become community priorities. These are the sources and chemicals that will be evaluated for possible community action.

Organize
Collect
Analyze
Mobilize

How does the Final Screen work?

The Initial and Secondary Screening steps provided an effective method to find and set aside many of the chemicals that have concentrations in community air that are at or below the community screening levels. In the Initial Screen, the Partnership used a simplified method combining readily available source information and a simple calculation based on the SCREEN3 air dispersion model to estimate concentrations. For the chemicals still needing further review after the Initial Screen, the Secondary Screen used readily available source information and the ISCST air dispersion model to get a more accurate estimation of concentrations. With many chemicals safely set aside by these screening steps, the Partnership will now focus its resources on a more detailed analysis of the chemicals still left in the screening process after the Secondary Screen.

All of the chemicals and sources under review in the Final Screen have been in the screening process from the start. The Final Screen will be the third time that the concentrations of these chemicals in community air will be estimated, each time using a more accurate method. For the Final Screen, the Partnership will review these chemicals by, once again, using the ISCST air dispersion model, only this time the Partnership will collect and use the best available source information as input for the model. To get this detailed information, the Partnership will organize itself to do things such as contact facility sources, work with transportation planners, or conduct surveys. The details of this information collection will be discussed below. Based on this effort, the Final Screen will give the Partnership estimates of concentrations in community air that are accurate enough and reliable enough to be used to identify community priorities.

As in the previous screening steps, these newly estimated concentrations will be compared to the screening-level concentrations that were developed in the Initial Screen. The chemicals with concentrations that are still above the community screening levels will become the community's priorities.

If a priority chemical has more than one source, the ISCST model will also be used to estimate the contribution that each source makes to the total concentration. This information will help the Partnership to target its efforts to the sources making the largest contributions to the total concentrations of the priority chemicals.

To complete the Final Screening step, the Partnership will be able to use the screening-level concentrations that were developed in the Initial Screen as well as all the work done in the Secondary Screen to set up the ISCST model with the community and source locations. At this point, the Partnership will only need to collect the best information on the sources and chemicals still left in the screening process and rerun the ISCST model using this new information.

What will be the results of the Final Screen?

The Final Screen completes the Partnership's effort to review all the chemicals released into community air and find those releases that result in concentrations estimated to be above the community screening levels. With the completion of this review, the Partnership will have two important pieces of information. First, the Partnership will have the list of the chemicals and the sources that will be identified as the community priorities. All the chemicals whose newly estimated concentrations are still above the community screening values will become the community priorities. And second, for each priority chemical, the Partnership will have an estimate of the contribution that each source makes to the total concentration. With these two pieces of information, the Partnership will be able to focus its efforts on the chemicals and the sources that have the greatest potential to adversely affect the health of members of the community. These chemicals are the highest priority for efforts to improve local air quality.

What new source information will the model need to produce the more accurate estimates for the Final Screening step?

Up to this point in the screening process, the Partnership has been using readily available information on the sources and releases in and around the Partnership area. For the Final Screen, the Partnership will contact, visit, and observe sources to collect and verify the best available information. This more accurate information will be used as input for the ISCST air dispersion modeling. The following is a list of the kinds of information that the Partnership will collect for each of the sources releasing the chemicals that are under review in the Final Screen.

For the Stationary Point Sources

- Release amounts: Instead of using maximum permitted amounts and other conservative overestimates, the Partnership will collect and verify estimates or measurements of actual release amounts from sources.

- Verified stack information: The Partnership will contact facilities to verify and, if necessary, correct the information on stack heights and locations collected from available databases for the Secondary Screen.

- Verified release characteristics: The Partnership will contact facilities to verify and, if necessary, correct the information on the release characteristics such as the velocity and temperature of the releases as they leave the stacks. This information was collected from available databases for the Secondary Screen.

In addition, the Partnership may want to consider the use of alternative methods, such as modeling fugitive sources as emissions released uniformly over a finite area rather than combining all fugitives into a single stack release. This can be easily accomplished using the area source modeling option in ISCST and information from the facility on the dimensions and chemicals released from on-site stationary fugitive sources.

For the Stationary Area Sources

In most cases, the concentrations developed in the Secondary Screen will be used again, and no further information on these sources will need to be collected. The Partnership will determine if any further refinement of the information used to estimate concentrations in the

Secondary Screen is available. For example, members of the Partnership may question the accuracy of the information used to estimate the concentrations due to home heating if the number of woodstoves with high emissions may have been underestimated. If the Partnership decides that the information used in the Secondary Screen could be significantly improved, then efforts to collect better information can be organized and new concentration estimates developed based on the refined information.

In addition, if the area source contributions are significant to the overall analysis, the Partnership may want to consider the use of an alternative to the pseudo-point modeling method used in the Secondary Screen. Stationary source releases can be modeled as emissions released uniformly over the area of the census tract rather than as combined releases of five pseudo-stacks as done in the Secondary Screen. If GIS resources are available, this can be accomplished easily using the area source modeling option in ISCST and information on census tract boundaries. Information for this approach is provided in the technical section of the Manual.

For Mobile On-Road Sources

At this point the Partnership will need to decide if further modeling will be necessary to identify mobile on-road chemicals as community priorities. Given the well-documented risk that these sources present in urban areas, the pseudo-point method used in the Secondary Screen may, despite its limitations, be sufficient to establish these chemicals and sources as community priorities. The Secondary Screen estimate can also be used to estimate the relative proportion that mobile sources contribute to the aggregate chemical concentrations in community air, keeping in mind that, due to the pseudo-point method used in the modeling, the concentrations at locations close to roads with heavy traffic may be higher. The Partnership could easily identify highways and streets with high traffic volume or idling vehicles by drawing on the residents' knowledge. With this information, the Partnership may be able to identify mobile sources as a priority and identify key neighborhoods and streets for special attention without the further use of air dispersion modeling to refine the concentration estimates.

Even if the Partnership decides that mobile sources can be identified as community priorities with no further modeling, to help find possible solutions the Partnership will need to collect as much detailed information as possible on the number and types of vehicles using the roads and highways targeted for possible community action.

If the Partnership finds that it needs more refined estimates of concentrations to develop recommendations or to get the support needed for actions to reduce exposures to mobile sources, the Partnership's technical team can use the ISCST model to develop more accurate estimations of concentrations near heavily traveled roads. The Partnership could also request monitoring in the areas most affected by mobile sources to directly measure mobile source concentrations.

To use the ISCST model to produce more accurate estimates of mobile on-road source concentrations in the community locations closest to the streets and highways with the most traffic, the Partnership's technical team will use information on vehicle miles traveled for each targeted street and estimate the emissions that will result from the vehicles using the street by using MOBILE6.2, an emissions model available from EPA's Office of Transportation and Air Quality. This model uses information on vehicle miles traveled to estimate highway emissions. The technical team will then use the ISCST air dispersion model to predict the dispersion of these emissions from the community's heavily traveled roadways and estimate the concentrations that will result in areas close to the modeled roads. Information for finding detailed guidance for using MOBILE6.2 and ISCST to estimate mobile on-road concentrations is provided in the Technical Guidance section of the Manual.

The information required to use MOBILE6.2 and ISCST to develop more accurate concentration estimates includes vehicle miles traveled, information on the key community highways and streets, gasoline formulations used in the area, temperatures, and other inputs.

For Mobile Non-Road Sources

As for non-road mobile sources, the concentrations estimated in the Secondary Screen will, in most cases, be adequate to determine if these sources will be targeted as community priorities and to estimate the contribution of these sources to the aggregate concentration from all sources. If the Partnership is aware of a local concentration of non-road sources, such as an airport, a railroad terminal, a port, or even a large construction site affecting its area, the Partnership may want to make a special effort to develop a more accurate estimate of the concentrations resulting from these sources. Methods to estimate emissions and concentrations from these sources have been developed. Please see Chapter 12, "Final Screen," in the Technical Guidance section for references to resources for estimating concentrations from these sources.

Trucks line up to enter a terminal at the Port of Oakland

How will the Partnership collect this new information?

Collecting the information listed above will require an effort to contact, visit, or observe the sources releasing the chemicals that are under review in the Final Screen. Depending on the chemicals and sources identified by the Secondary Screen, the Partnership may need to survey traffic on major roads or contact and visit both large and small commercial, industrial, and public facilities. To make these contacts and collect the information listed above, it may be helpful for the Partnership to set up teams and divide up the responsibility for collecting information on the different sources. It will help if most of the representatives from the stationary sources are already participating in the Partnership. These members can facilitate the collection of the information that will be needed from the facilities they represent. If a facility source is not represented in the Partnership, the team will need to contact the source and introduce the Partnership to collect the information that will be needed.

Fenix Barbour, a resident of West Oakland, monitors truck traffic along 7th Street in West Oakland.

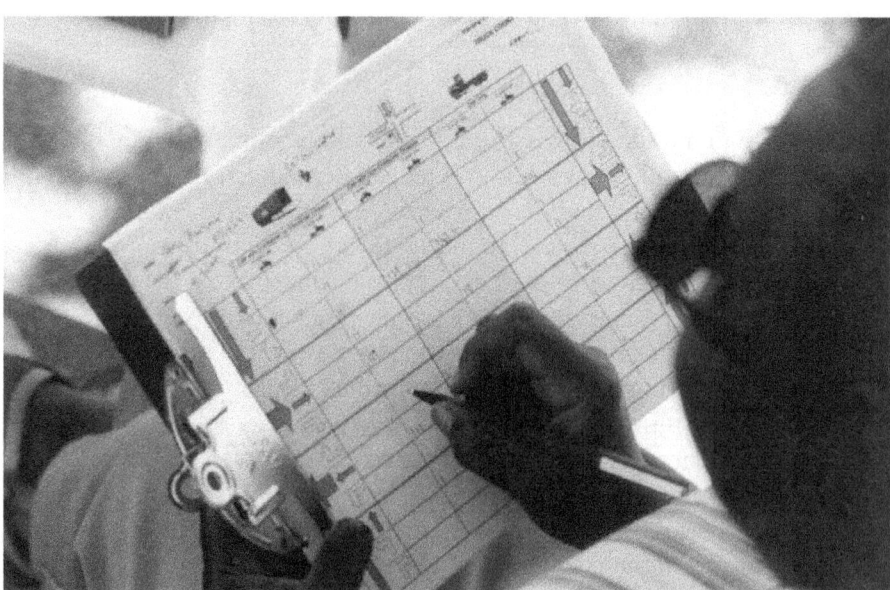

Photos on this page courtesy of Pacific Institute

If necessary, collecting more detailed information on the releases from the on-road mobile sources may require contacting local, regional, state, and tribal transportation experts and getting their help to collect the additional information on traffic that will be needed. The Partnership may also need to mobilize community resources to conduct traffic surveys if the information needed is not available. Local schools or community organizations may be able to help with the collection of this information.

In addition to collecting information, the teams will need to develop a method for verifying the accuracy of the information that is collected. For example, the team could review the methods used for estimating releases from stationary sources to ensure that they are reliable or review production and chemical use information to ensure that all releases are accounted for.

The Partnership teams set up to collect the information may be most effective if they contain a cross section of Partnership members. Community leaders and local residents will be key to explaining the work of the Partnership and convincing the facility sources to cooperate and join in the effort. Technical members of the Partnership will be needed to help with the collection and verification of the information. The effort to contact and collect information from the facility sources reviewed in the Final Screen will provide an excellent opportunity for the members of the Partnership to get to know these sources and to strengthen the Partnership.

How will the Partnership complete the modeling for the Final Screen?

Once the Partnership teams have collected the new and more accurate information on the sources and releases, the technical team responsible for modeling will use this information and the ISCST air dispersion model to more accurately estimate concentrations for the chemicals under review. As in the Secondary Screen, the model will be set to estimate air concentrations at the locations designated by the Partnership. The model will also identify all locations in the community with concentrations that are estimated to be above the community screening levels. If the concentration estimates for stationary area and mobile sources developed in the Secondary Screen are used for the Final Screen, please remember the limitations to estimating concentrations at specific community locations, discussed in the previous chapter, that result from the pseudo-point method that was used to estimate the concentrations from these sources.

As the priority chemicals are identified, the technical team will also use the ISCST model to determine the contribution that each source makes to the total concentration. Guidance to help the technical team run the ISCST model to estimate concentrations and the contribution of each source can be found in the Technical Guidance section of the Manual.

What are all the steps the Partnership will need to complete the Final Screen?

Please see the following page for a list of all the steps the Partnership will need to complete during the Final Screen.

Once the Final Screen has been completed and the community's priority chemicals and sources have been identified, it will be important to report this information to the community and develop recommendations for possible actions to reduce exposures. These important Partnership activities will be discussed in Chapter 8.

What are the key decisions the Partnership will need to make or monitor to ensure that the screening continues to meet its needs?

At this point, the oversight role of the Partnership will probably need to focus on the review and approval of the methods that the teams will use to collect and verify data. The full Partnership will need to participate in these decisions since the chemicals identified in this step will become the community's priorities, and everyone will need to be confident that the information used to estimate concentrations was reliable and accurate.

The Partnership may also need to decide on a policy for estimating concentrations if sources are not willing to provide, and the Partnership cannot find, the information that it needs to improve the modeling estimates.

What skills and resources will be needed during the Final Screening step?

The Partnership will, once again, need to draw on the resources of most of its members. In addition to the technical skills needed to run the ISCST air dispersion model, the Partnership will need the following skills.

Summary of Steps to Complete the Final Screen

Step 1: Provide background education to the full Partnership on the kinds of information that will be needed to improve the estimations of releases and education on the possible methods that the Partnership can use to verify the accuracy of this information.

Step 2: Agree on the kinds of information that will be collected for the Final Screen and on the methods that will be used to verify the accuracy of the information collected.

Step 3: Set up teams and divide responsibilities for contacting the stationary sources and collecting information on the mobile sources that release the chemicals reviewed in the Final Screen.

Step 4: Set the quality control system to check the data collection and analysis to ensure that it follows Partnership guidelines and that it is accurate and complete.

Step 5: Collect information by the best means possible including, if necessary, visits to facilities, traffic surveys, etc.

Step 6: Run the ISCST model using new information to develop new concentration estimates.

Step 7: Compare new estimates to community screening levels and identify chemicals with concentrations that are still above the screening levels as the community's priorities.

Step 8: Identify the sources of the priority chemicals and run the ISCST model to estimate, for each chemical, the contribution of each source to the total concentration.

Technical skills: To collect and verify the new emissions and source data, the Partnership will need members familiar with methods used to estimate releases from various sources and with methods that can be used to verify release estimates. Members familiar with modeling and the information used in models will also be necessary to help determine the kinds of information that the Partnership will collect. University, industry, and government technical staff and retired community professionals will have these skills.

Science education skills: To participate in or monitor the collection and verification of the new information that will be used in the Final Screen, the full Partnership will need background information and education on kinds of information used for air dispersion modeling and on the methods that can be used to verify the accuracy of information collected from or about sources. Committee members with skills in communicating technical information will be needed to provide this training to the full Partnership. Science teachers from local schools, colleges, and universities, and technical members of the Partnership with experience in communicating technical information, should have the skills needed to provide this training.

Communication and partnership skills: Contacting sources, explaining the Partnership, and encouraging representative of sources to provide information and join in the Partnership effort to improve air quality will require communication and partnering skills. Any member with experience in communication will be able to participate in this effort.

How can the work of the Final Screen be carried out in a way that builds the long-term capacity of the community?

As in the previous steps of the screening, the Partnership's technical team can provide training to local residents, teachers, and students and help them to run the ISCST model. This approach to the work may take more time, but it would help to build the long-term capacity of the community and the Partnership and help to ensure that all the skills needed for future reviews of community air will be available in the community.

What are the next steps?

In addition to learning about the sources and releases of the chemicals in community air and identifying priorities, the purpose of the Partnership's work is to mobilize the community to reduce risks from chemical releases. As a result, the Partnership's work does not end when the results of the Final Screen come in. Once the screening is completed and priorities have been identified, the Partnership will take the information learned through the screening process and use it to inform the broader community, develop recommendations for possible actions that can be taken to reduce exposures, and mobilize the community to take action where possible. These next steps will be discussed in the next chapter.

Overview

Communicating the Screening Results and Developing Recommendations

What will the Partnership need to do to use the results of the screening analysis to achieve the Partnership's goals?

It is important to remember that the technical screening work that has been the main focus for the Partnership for some time is not an end in itself. The screening effort and its results are a means designed to help the Partnership reach its overall goals. The overall goals of the Partnership were discussed and set at the beginning of the effort, and the next steps for the Partnership will be determined by these goals. Partnership goals may include things such as improving local air quality, or developing a better understanding of local air quality and local sources of pollution, or building the long-term capacity of the community to understand and improve local air quality, or some combination of these or other goals. Whatever goals have been set, to reach them the Partnership will, almost certainly, need to work with and mobilize broad sections of the community to make further progress. As a result, the Partnership's emphasis will now shift from primarily studying air quality to the task of informing and mobilizing the community to take actions needed to reach its overall goals.

Organize
Collect
Analyze
Mobilize

The Partnership's emphasis will now shift from primarily studying air quality to the task of informing and mobilizing the community to take action

Once the screening analysis is completed, what steps will the Partnership need to take to reach its goals?

The following is a list of the steps the Partnership will need to take once the screening analysis has been completed.

Summary of Steps to Reach Partnership Goals

Step 1: Prepare a report and communication materials summarizing the work and the results of the screening process.

Step 2: Develop, if possible, feasible recommendations for possible actions the community can take to effectively reduce exposures to the priority chemicals and sources.

Step 3: Communicate the results and the recommendations to the broader community.

Step 4: Expand the Partnership so that it has all the resources necessary to mobilize the community to take actions that will improve local air quality.

What skills, resources, and organization will the Partnership need to prepare a summary report, develop recommendations, and communicate the results and recommendations to the broader community?

All the members of the Partnership will need to contribute their time and skills to move the Partnership's work from study to action. Skills needed will include the following.

Communication skills: Preparing the summary report and communicating the report and recommendations to the broader community will require a broad range of communication skills, from writing and art to presentations. A good understanding of community questions and views on air quality will also be essential to good communication. Partnership members who know the community can team up with members who have experience in writing and communicating to explain the results of the screening and the recommendations. Teachers, reporters, students, technical staff with experience in communication, and community members and leaders can provide these skills.

Community leadership and organizational skills: Using the screening results and recommendations to organize and mobilize the community to take action will require leadership and organizational skills from all sectors of the community. Business leaders will be needed to mobilize their colleagues to carry out recommendations relating to business. Similarly, community, church, and school leaders will be needed to organize and mobilize the community to address other community priorities and sources. Community and business leaders who are members of the Partnership will have these skills, and additional leaders can be recruited to the effort as the work progresses.

Technical and research skills: Developing recommendations for reducing exposures to priority chemicals and sources may require a broad range of technical skills. Recommendations, depending on the chemicals and sources, could include things such as changes in chemical uses in industrial processes or in community households. Technical staff from local industry and business or from their trade associations, government technical and pollution prevention staff, and college and university staff will have the technical background and skill to develop recommendations. Research skills will also be important to help find technologies and approaches that have been developed and used by the similar pollution sources in other communities.

Creativity: Finding effective and practical ways to reduce exposures will probably require the Partnership and community to come up with some new and creative approaches. The work to improve air quality at the local level is new, so there are not a lot of ready-made solutions for the Partnership to use. The Partnership will need to encourage and draw on all of its members to find the solutions it will need.

How will a report summarizing the work and the results of the screening analysis support the Partnership's efforts to inform and mobilize the community to take action?

A written report summarizing the work done to study community air and to identify the chemicals and sources that will be community priorities will be a tool for the Partnership to use to build the consensus in the community that will be needed to take action. Many members of the community will not have the opportunity to participate directly in the work of the Partnership. The report will be key to providing these members of the community with all the information that they will need to understand and join in the work to improve local air quality. This information and the open, thorough, and objective screening method that the Partnership used to identify the priority chemicals and sources will provide the basis for building a consensus in the community on a plan for improving air quality. A clear presentation of the limits of the information in the report will also ensure that community members understand the information and can place it in context with other information. With all sectors of the community united and mobilized around a plan for possible action, the community will have the best possible chance of achieving its goals. As the saying goes, "Where there's a will, there's a way," and the Partnership's report will be key to creating the "will" in all sectors of the community.

The development of a report summarizing the results of the screening will also be a key opportunity for the Partnership to build the community's capacity to understand and improve local air quality. The Partnership can use the community's interest in the priority chemicals and sources as an opportunity to share the background information on air quality that the community will need to participate fully in the work to improve air quality. To accomplish this, the report will need to be written in a clear and understandable fashion. Some methods the Partnership might use to help ensure that the report communicates effectively will be discussed in the next section.

To meet the needs of the community, the Partnership will probably need to produce several different versions of its report, including a short summary in the form of a press release, a short summary designed to be widely distributed and read by almost everyone in the community, and a longer report of the work of the Partnership for those in the community who will be interested in a detailed account of the work. All of these versions will help to build the broad consensus that will be needed to take possible action.

While a written summary report of the Partnership's work will be crucial, it is important to remember that this report will not be the only means the Partnership will use to communicate the work it has done. To reach its goals, the Partnership will need to use a variety of different means to communicate with the broader community. A discussion of how the Partnership might conduct a broad campaign to communicate the results of the screening and the recommendations it has developed is discussed below.

What methods can the Partnership use to develop a report for the community that effectively summarizes the results of the screening process?

Developing and maintaining a consensus as the work progresses in the Partnership, and communicating with and getting input from the community at each stage of the screening analysis, are the most effective ways to prepare the Partnership to write a final report that can effectively communicate the work to the community. If the Partnership has communicated consistently with the community throughout the screening work, it will have the experience needed to effectively communicate the results of the analysis.

Once the screening analysis is complete, the first step in developing a summary report will be to thoroughly discuss the results of the screening and what they mean for the community in the full Partnership committee. Before starting to write a report, the Partnership should reach a consensus on all the key points that will be made in the summary. This consensus in the Partnership will provide the basis for beginning work on the report. The time needed for the Partnership to reach a consensus on the key points of the report will depend in part on the amount of time the Partnership took to clarify its thinking at each stage of the work.

An effective report will directly address the concerns of the community. The community will look to the report for answers to their questions and concerns about local air quality. Understanding these questions and concerns, and addressing them directly, will be key to developing an effective report. The screening analysis will probably not answer all community questions, and it will be important for the report to point out which community questions can and can't be answered with the work that the Partnership has done. Once again, practice at communicating the work of the

Partnership and paying attention to community input are the best ways to prepare for understanding and addressing community questions and concerns.

To help with drafting its summary report, the Partnership team that is given responsibility for drafting the summary report may want to involve community members, both residents and business representatives, who have not participated in the previous work of the Partnership. Their participation would help to ensure that the report is written in a form that is understandable to everyone in the community and not just those who have participated in the Partnership's work.

Once the Partnership has agreed on the form for the summary of its work, either a written summary of the main points or a draft version of the full report, it would be helpful for members of the Partnership to practice presenting these ideas to small groups in the community. This will give the Partnership a chance to identify any questions that have not been addressed or areas that need better explanation. Input from these small meetings can then be used to complete the final report.

It will also be important for the Partnership, even after it has prepared and presented its final report to the community, to remain open to input from the community and to be willing to address important community issues that may have been missed.

Why should the Partnership consider working on developing recommendations for reducing the exposures to the priority chemicals and sources?

Developing and presenting recommendations for potential actions that can be taken to reduce exposures as soon as the priority chemicals and sources are identified will help to ensure that the community gets a quick start on its effort to improve community air quality. In addition, presenting possible solutions side by side with the priorities identified by the screening process may help to relieve community concerns and channel community energies into the efforts that will be needed to address those concerns. If facility sources are identified as community priorities, developing recommendations for reducing exposures will also give these facilities an opportunity to demonstrate their willingness to work with the community to address concerns and improve air quality.

Should the results of the screening analysis and the recommendations for addressing the priority chemicals and sources be communicated to the community at the same time?

Each Partnership will have to determine the most effective way to develop recommendations and communicate the results of the screening for their community. The following are considerations to keep in mind while planning this work. First, the Partnership will probably want to communicate the results of the screening to the community as soon as a clear summary can be written. To delay in communicating the results could contradict the Partnership's goal of informing the community and possibly undermine the community's trust in the Partnership. If the Partnership delays releasing the results of the screening so that it has time to develop recommendations, the community may feel as though important information has been kept from them, especially if there are actions the community members could take immediately to reduce risks. And, given the broad nature of the Partnership, the community is certain to learn about the results of the screening even if the Partnership delays its report while it develops recommendations.

Once the Partnership reports the results of the screening process to the community, it should make every effort to develop feasible recommendations for addressing the priority chemicals as soon as possible. Presenting recommendations for possible actions that can be taken to reduce risks from priority chemicals and sources as soon as possible will help to alleviate community concerns and shorten the time needed to reach the goal of improving air quality. Any long delay between the identification of the priority chemicals and the actions to address them is likely to increase community concerns.

It may be possible to identify in advance some chemicals that are likely to be priorities, and the Partnership, if it has sufficient resources, could organize a team to begin developing recommendations even before the screening analysis is completed. Mobile source chemicals with high concentrations throughout the nation could, for example, be selected for early work to develop recommendations. On the other hand, some chemicals will not be obvious priorities and, once they are identified, may require considerable time for the Partnership and community to develop an effective and practical plan to reduce exposures.

As an illustration, the following is an example of a plan the Partnership could adopt to try to meet community needs at this stage of its work: As soon as the results of the screening are in, the Partnership develops a clear written report summarizing the results. The Partnership distributes this report to the local press and the community along with an announcement letting the community know that the Partnership will be working on ideas for addressing the priority chemicals and sources and that there will be open community meetings in a month to discuss both the results and the recommendations for addressing the concerns. The Partnership then spends a month working to identify as many effective and practical recommendations as possible. Chemicals and sources that need additional time and community input to develop effective plans for action are also identified, and further work to develop a plan for reducing risks for these chemicals and sources are included in the recommendations. At the community meeting following this month's work, the Partnership presents the results and the preliminary recommendations for action as the beginning of a major effort to communicate with and mobilize the community.

What should the Partnership consider to develop recommendations and to decide which ones should be implemented?

To develop and choose recommendations for implementation, the Partnership will need to carefully discuss all options to find the ones that best match community goals. The Partnership may, for example, have more recommendations than can be implemented at once with available resources. In this case, the recommendations will need to be prioritized and ranked in terms of their ability to achieve community goals. It will be especially important at this point for the Partnership to take time to review the discussion of goals held at the beginning of the Partnership's work and to clarify and update these goals so that they reflect everyone's current thinking. Once the goals of the Partnership and community have been clarified, recommendations can be discussed with these goals in mind. For example, if the Partnership's goals are to reduce risk from air toxics, improve community health, and build community capacity, then in order to develop and choose among recommendations, the Partnership will need to consider things such as the following: the amount of risk reduction that a recommendation will achieve in comparison to other recommendations, the cost and the time it will take to implement a recommendation, and the contribution of a recommendation or a combination of recommendations to strengthening the Partnership and building the enthusiasm of all sectors of the community to contribute to risk reductions. Partnerships may also want to consider community perceptions of risk and the relationship of the recommendations to these perceptions.

The relationship of the recommendations to other community goals should also be considered. For example, it will be important for the Partnership to take time to consider the impact of work to improve air quality on the other ongoing efforts to improve community health. This will enable the Partnership to design the work to improve air quality in a way that complements and supports the other work to improve community health. In most cases, this work to coordinate with other community efforts will result in more effective work to improve air quality. In some cases, especially in communities with limited resources, a partnership may decide, based on the information obtained from the screening analysis, that other community concerns, such as contaminated food or drinking water, are more pressing concerns for the community. As a result, they may delay implementation of the recommendations on air toxics until these risks are addressed.

Finally, although the goal of the screening exercise is to identify community priorities so that effective efforts to reduce risk can be started, in some cases a Partnership decision to recommend further study of the priority chemicals and sources may be warranted. For example, the Partnership may decide to seek assistance for a monitoring study to verify the results of the modeling. This may be necessary to resolve any doubts in the Partnership about the data used for modeling or the accuracy of the modeling. The Partnership could also decide to begin reduction actions based on the modeling results and carry out a monitoring study at the same time. Or the Partnership could decide to request assistance from state or federal agencies to conduct a more in-depth exposure investigation of the priority chemicals so that information on risk can be used to help make decisions on recommendations and possible actions.

What should the Partnership consider to develop recommendations for each of the priority sources and chemicals?

To do the work to develop recommendations for each of the priority chemicals and sources, the Partnership may need to organize a team for each source or source type. The members of the teams that collected information on the priority sources for the Final Screen will be good candidates for membership on these teams since they will already be familiar with the sources. For example, the Partnership might organize teams to develop recommendations for mobile sources, for auto body shops, for households, for a refinery, for an incinerator, etc. The following is a list of suggestions for these teams to consider in their work to develop recommendations.

Relying on the community, residents, and businesses to find practical solutions: Whatever the chemical or source of concern, to find solutions that will work and can be implemented, the Partnership will need to rely on those affected to help develop the recommendations. Businesses and residents who understand and support the goals of the Partnership will have the best chance of finding ways to change business practices and community habits. The Partnership may need to recruit new members to help with the development of recommendations for reducing emissions of the priority chemicals. Representatives of all the major sources, if they are not already members of the Partnership, should be recruited to participate in the effort to find solutions.

Looking for models and benchmarks: In addition to drawing on its members and the community for ideas on reducing emissions and exposures, the Partnership should also look for information on efforts to address similar chemicals and sources that have been present in other communities. Successful programs and technologies may have been developed that the Partnership can adopt or learn from. Businesses in other communities that have developed model emission reduction programs can be used as benchmarks and as a source for ideas. Similarly, successful community efforts to address mobile and household sources can be a source for ideas.

Using a pollution prevention approach to address the priorities: When possible, the Partnership should look for pollution prevention opportunities in the recommendations it develops. Pollution prevention focuses on reducing the release of pollutants by limiting or eliminating the use of toxics at the source, instead of looking for ways to control the releases and exposures that result once the toxics are used. Pollution prevention solutions eliminate the need for ongoing management and reduce the possibility of exposure from accidental releases. In addition, many businesses have found that pollution prevention can help to cut costs.

Applying for funding: As a part of its plans, the Partnership may also need to consider applying for funding to support its work. There are a variety of funding sources designed to support community efforts to reduce emissions, and the Partnership, based on the work it has done

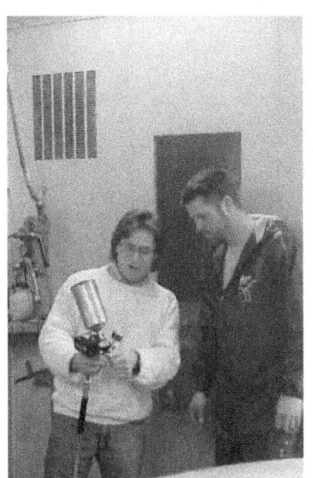

Jeff Aigeldinger calibrates a painter's spray gun and offers advice during a site visit to an auto refinishing shop. Best practice information for auto refinishing shops was developed by EPA's Design for Environment Program

to set priorities and mobilize local resources, will be in an excellent position to qualify for this funding. To take advantage of current and future funding opportunities, the Partnership may need to organize training to build the capacity of the community to apply for funding. Some members of the Partnership should be able to provide this training or know where it can be found.

Providing incentives: It will also be important for the Partnership to include in its recommendations ways to provide incentives to businesses and residents for reducing emissions. In the long term, the incentive of building a healthy and sustainable community will have the most impact, but finding ways, in the short term, to recognize and reward efforts to support community goals can be very effective. Developing community "green" or "clean" business, household, or school recognition programs may be very helpful. Models for developing these kinds of incentive programs are available.

Using enforcement as an option: While, in most cases, the potential actions required to meet community goals will go beyond regulatory requirements and require voluntary efforts, if there are violations of environmental regulations and the Partnership cannot convince a business to join in community efforts, the Partnership recommendation may call on its government partners to use their enforcement authority to address community concerns.

Partnering with nearby communities: In the cases where the Partnership will need to address sources that have regional impacts, such as mobile sources or large facilities, the Partnership may find that it needs to join with other nearby communities to develop a plan that can effectively reduce emissions. Because of the need to consult with other communities, recommendations to address these concerns will probably take longer to develop and implement. Despite the time required, the Partnership is likely to find that some of the community's key concerns can only be addressed by reaching beyond its own borders and working with other communities.

Developing both long- and short-term recommendations:
Combining recommendations that can produce results in a short amount of time with recommendations that will take longer to develop and implement may help to sustain community involvement. Making and publicizing short-term accomplishments will build community confidence and enthusiasm and help to maintain the efforts that will be needed to address larger or more complex concerns. For example, changing to low-sulfur fuel for local buses, getting facilities to commit to early reductions in emissions, or

working with local retailers to reduce household chemical use can produce measurable results in a year or less, while the planning and work for longer-term projects, which may take two or more years, can be developed, such as redirecting truck traffic or developing and providing pollution prevention training to small local businesses.

Recommending pollution prevention and emission reductions as community standard for chemical use: In addition to the key task of developing recommendations to address the community's priority chemicals and sources, the Partnership should consider developing a general recommendation and incentive program that encourages continuous emission reductions and pollution prevention for all chemicals used in the community. This is especially important given the limited information that is available to communities to conduct screening exercises. As the Partnership learned, many chemicals do not have adequate testing data, and new testing may discover additional hazards for even well-tested chemicals. In light of these uncertainties, a recommendation that supports and encourages a continuous commitment to reductions and pollution prevention for all community chemicals would complement the specific recommendations developed for the priority chemicals.

Please see Appendix G for a list of resources and programs for pollution prevention and emissions reduction.

How can the Partnership organize a campaign to effectively communicate the results and recommendations and mobilize the community to potentially take action?

Once the summary report and preliminary recommendations have been developed, the Partnership will probably need to focus for a time on a major effort to communicate this information and mobilize the community to take potential action. All the members of the Partnership will need to be involved in this effort to communicate with the community.

To develop an effective outreach campaign, it will be important for the Partnership to clarify its goals and message so that all the members of the Partnership are speaking with one voice in the community. Goals for the outreach might include mobilizing the community to take some possible action or actions, improving the community's understanding of local air quality, and recruiting volunteers to implement the next steps of the Partnership's work. The Partnership should also agree on the key points from the screening

All the members of the Partnership will need to be involved in the effort to communicate with the community

analysis and the summary report that it wants to stress and prepare presentation materials summarizing the goals and message that everyone in the Partnership can use. The Partnership will also need to plan to discuss and respond to any input that the community provides during the course of the outreach.

With a common understanding of the message and goals, the Partnership can organize small outreach teams or assign individuals to represent the Partnership in local media and to speak to community organizations such as block clubs, PTAs, business organizations, schools, and churches. Teams or individuals can also meet with key community leaders and with key stakeholders outside the community. Combining Partnership members from different sectors of the community on the same outreach team will improve the effectiveness of the outreach and provide an excellent opportunity for strengthening the Partnership. The Partnership may also want to organize its own public community meetings to communicate with and mobilize the community for the next steps.

Will the Partnership be able to find the community resources that will be needed to develop and implement the recommendations for improving air quality?

The campaign to communicate the results of the screening and the recommendations to the community will provide an excellent opportunity for the Partnership to recruit the resources it will need to reach its goals. The communication effort to explain the priority chemicals and sources and the work done to identify them will help to develop community interest in local air quality. The recommendations for changes that can improve air quality will also be sure to attract the attention of the community, especially those directly affected by any proposed changes. Small businesses that are the focus of a recommendation will be encouraged to participate by the prospect of getting help finding the resources and information that they need to protect their workers and neighbors. Residents will actively help with measures designed to improve community health by reducing exposures from the use of household chemicals or from mobile sources.

The Partnership itself and its work will also serve as an incentive for greater community participation. As the outreach campaign unfolds and teams from the Partnership meet with groups throughout the

The campaign to communicate the results of the screening and the recommendations to the community will provide an excellent opportunity for the Partnership to recruit the resources it will need to meet its goals

Partnership area, the community will learn about the successful completion of the screening work and see a first-hand demonstration of the community's ability to work together to get things done. The Partnership's successes may be the most effective encouragement for broader community participation.

The Partnership's switch from a focus on understanding air quality to possible actions to improve air quality will also bolster participation. Efforts focused on making concrete measurable improvements in community air will attract many community members who were not interested in the more analytical aspects of the Partnership's work.

Overall, the opportunity that the Partnership offers for the community to focus on the chemicals and sources that have the greatest potential to adversely affect the health of community members and to make real improvements in community air will ensure the participation needed to move forward. To make sure that the new community volunteers do not waste their time, the Partnership will need to set up and organize new committees or other forms of organization prepared to focus on the implementation of the recommendations. These new committees will become the center of Partnership activity as the Partnership shifts to the work to improve air quality.

What can the Partnership do to ensure that the community's work to improve air quality will continue in the future?

At some point the Partnership will need to take time to consider and plan for the long-term work that will be needed to monitor and continue to improve local air quality. The work to develop and implement recommendations will also take time and planning, and long-term goals may need to be set for this work. Plans for updating the Emission Source Inventory database annually and plans for the next thorough screening of local air quality will need to be considered. The ability to sustain the work in the long term will also require the Partnership to plan for long-term funding and for the maintenance of some form of organization to support the work. The shape of the organization the Partnership will take will differ from community to community depending, in part, on the Partnership's relationship to other organizations in the community. The Partnership will also need to think about developing the leadership and providing the training that the community will need to carry on this work in the future. Keep in mind that the best way to ensure the long-term viability of the work may be to take advantage of the current work to improve air quality as an opportunity to build community capacity. Special training for community leaders and for local teachers could also be set up on a regular basis. For example, the Partnership could help teachers incorporate the science and knowledge needed to understand local air quality in the curricula of local schools so that the community's students are prepared with the background they will need to continue and to lead the work.

What can the Partnership do to share its experiences with other communities?

Communities across the nation are working to understand and improve local air quality. All of these local community efforts would benefit immensely if communities found ways to share their experiences and learn from each other. Sharing of information will also help to identify the air quality problems where the joint action of many communities and governments might be helpful or necessary.

Local community efforts would benefit immensely if communities found ways to share their experiences and learn from each other

To share information with other communities, the Partnership will need to assign one of its teams to take on this responsibility and provide them with the resources they will need. Setting up an Internet site for the Partnership and making all Partnership reports, decision documents, and outreach materials available on the web site would be an effective way to make the Partnership's experiences accessible to other communities. Including regular summaries of lessons learned on the web site would be especially useful. And an opportunity, such as a partnership with a local school with video training capacity to record the community's experiences in a video documentary, would be an excellent way to share experiences with other communities.

In addition, national meetings can be organized to bring communities together to share their experiences. If possible, the Partnership should plan to send members to participate in these meetings. At some point, the Partnership may want to take the lead to facilitate the sharing of information. For example, a Partnership committee working on developing a recommendation for a particular source may find it useful to organize a national conference call with other communities working on the same concern. As work at the local level develops, additional opportunities for exchanging information will develop.

Building the Emission Source Inventory

This section describes the procedure the Partnership's Emission Source Inventory Technical Team will use to set up the Emission Source Inventory database needed for the screening process. It includes the software and hardware requirements, a description of the information to be collected, a list of the steps that need to be completed to set up the database and collect the information, and detailed suggestions to help the technical team carry out each task. The information in this section is based on experiences of the Baltimore Community Environmental Partnership and on the comments received during the review of the technical report summarizing these experiences (refer to Baltimore Community Environmental Partnership Air Committee Technical Report, EPA 744-R-00-005, April 2000).

What is the overall goal for the technical team?

Chapter 4 of the Overview provides a general introduction to and background information for building the Emission Source Inventory. The overall goal for the technical team is to set up the Emission Source Inventory database and enter information needed to complete the Initial, Secondary, and Final Screening steps. A flow chart showing the overall procedure for developing the emission source inventory is shown in Figure 9-1.

The technical team begins the process of building the Emission Source Inventory database by collecting and entering release and location data for all the stationary and mobile sources in the study area. Information on background and relevant monitored concentrations are also entered in the database at this time. Detailed guidance for setting up the Emission Source Inventory database, identifying and locating sources, collecting release information, and collecting background and relevant monitored concentrations is provided in this section of the Technical Guidance. The process of collecting information about emission sources and

entering this information into the database continues through the Final Screening step. At each step of the screening process, the Emission Source Technical Team will collect the additional information that will be needed to estimate ambient air concentrations. Guidance for collecting the additional information that will be needed for each step of the screening process is provided in subsequent sections of the Technical Guidance.

What software and hardware will the Partnership use to store the emission source information needed for the screening process?

The software needed to create the source inventory database can be either a spreadsheet program (e.g., Lotus or Excel) or a database management program (e.g., dBase or Oracle). A database program may be preferable if your study area includes a large number of emission sources and chemicals, but in most cases a spreadsheet program will be sufficient. The software program is used to store, organize, and manipulate the data collected by the Partnership. The software must be able to query, sort, and perform mathematical manipulations. In addition, the software must also be able to export selected information from the Emission Source Inventory database for further analysis using air dispersion modeling software during the Secondary and Final Screening steps.

The hardware requirements include a personal computer, modem, and printer. The computer will need appropriate processor speed, random access memory (RAM), and hard disk space to run the spreadsheet or database software and the air dispersion modeling software. In addition, an Internet service provider (ISP) connection will need to be established to enable the Partnership to send and receive e-mail, perform Internet searches, and download data from various federal, state, and local government web sites.

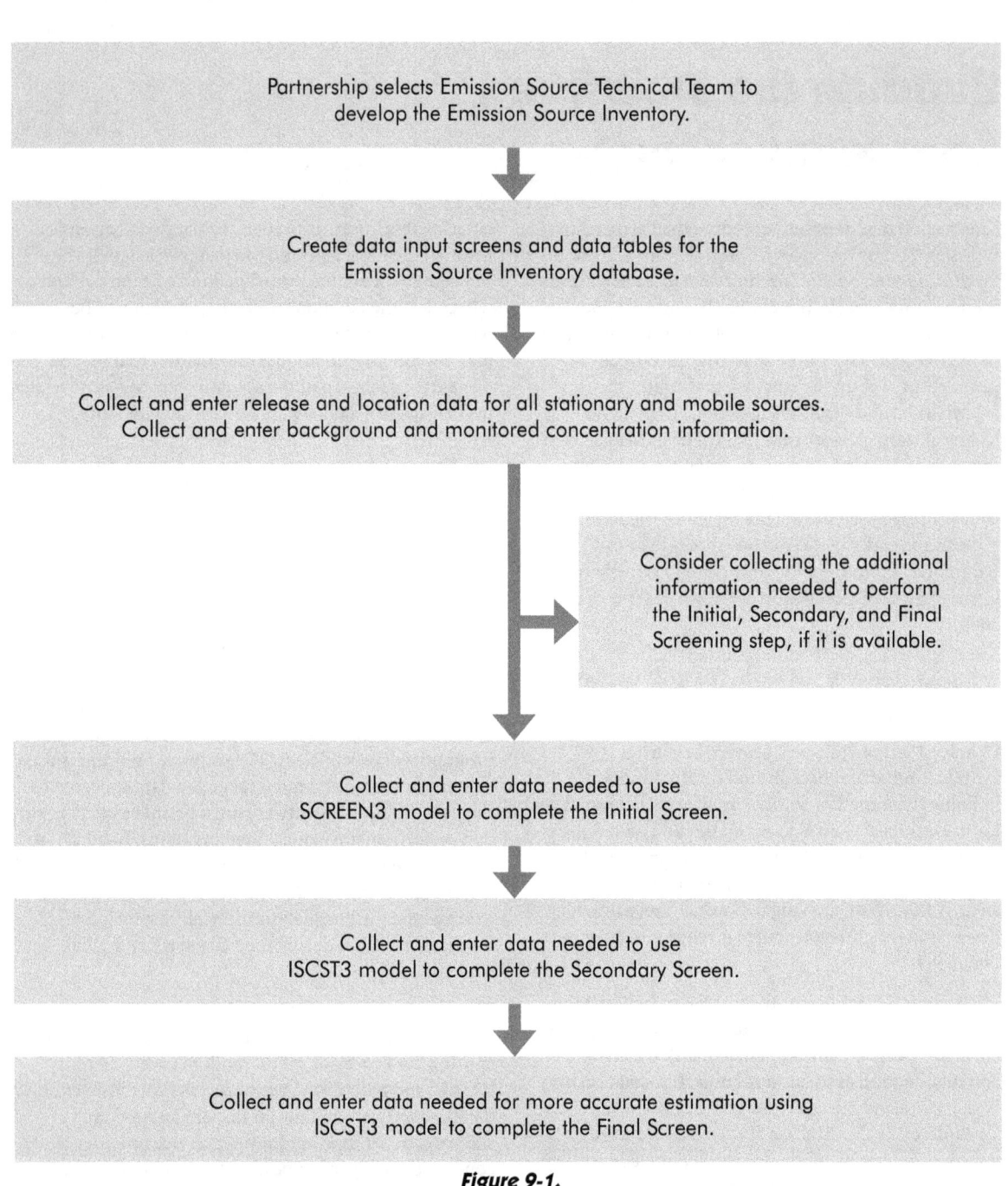

Figure 9-1.
Overall Procedure for Developing Emission Source Inventory

What information will be collected and entered into the database to begin developing the Emission Source Inventory?

Once the Partnership has defined the boundaries of the study area, data on source locations and releases are collected and entered into the Emission Source Inventory database. Keep in mind that emission sources outside the study area may also have an impact on air quality within the study area and, therefore, consideration should be given to including these sources in the Emission Source Inventory. Chapter 4 of the Overview describes the procedure for defining the boundaries of the study area. The goal is to capture information on as many chemicals and emission sources as possible in the study area. The technical team will need to collect information from several databases or information sources maintained by different government agencies. Each of these databases and data sources contains information on selected chemicals. Many of the same chemicals will be in all the data sources, but some chemicals will be found in one database but not in other databases. The Partnership will use all available data sources to collect information on as many chemicals released in its area as possible. Detailed guidance for accessing this information is provided below.

Table 9-1 summarizes the information that needs to be collected for each source type prior to the Initial Screen. The focus at this point is to collect release and location information for all sources, as well as background and relevant monitored concentration data. But, if additional information needed for the Initial, Secondary, and Final Screens is readily available, then consideration should be given to collecting this information too.

During the Initial and Secondary Screening steps, the Partnership will be using readily available release data based on maximum permitted release amounts as inputs for the look-up table and the ISCST model, respectively. If the release data available to a community Partnership have only estimated actual release amounts and not maximum permitted release amounts for stationary sources, the Partnership will need to increase these release estimates, possibly multiplying the release amount by a factor agreed on by the Partnership, to ensure that the Initial and Secondary Screens are conservative. For example, release estimates taken from the TRI database are estimated actual releases, not maximum permitted releases, so the Partnership's

technical team will need to increase these amounts to ensure that the Initial and Secondary Screens are conservative. For the Final Screening step, the Partnership will have the resources to contact each of the remaining stationary sources to obtain actual release data.

A table summarizing the information needed to complete all of the screening steps is provided at the end of the Overview section for "Building the Emission Source Inventory," Chapter 4. The decision to collect additional information needed to complete the Secondary and Final Screening steps should take into account the fact that the Partnership will not know in advance which chemicals will be identified for further screening during each step of the process. For example, detailed information about the temperature of the gases released from stationary point sources (i.e., stacks) is needed to estimate concentrations during the Final Screen. But collecting this information for all the stack releases in the Emission Source Inventory would be a waste of time, since the information will only be needed for the small number of chemicals left after the Secondary Screen is completed.

Despite the fact that the Inventory Technical Team will need to wait until the screening results are complete to know which chemicals will remain in the screening process, it is important for the Partnership to know all the data that will be needed to complete the screening. Understanding what the data requirements are allows the Partnership to take advantage of easy opportunities to collect information as they arise. For example, if the detailed information on releases needed for the Final Screen is present in the same database used to collect the information on the amount of releases, and if the information is in a form that will allow it to be transferred electronically into the Emission Source Inventory, then detailed information on all the chemicals can be collected from the beginning with minimal effort.

Table 9-1.
Information Collected for Emission Source Inventory Prior to Initial Screen

TYPES OF SOURCES	INFORMATION TO BE COLLECTED PRIOR TO THE INITIAL SCREEN
Stationary Point Sources: All large and small commercial, industrial, and publicly owned facility sources (sources will be modeled individually)	
• Sources with individual release information available	• Facility name • Facility location • Chemical released • CAS number • Emission rate (lb/yr)
• Sources with individual release information **not** available (sources will need to be estimated)	• Facility name, location, chemical released • Emission factor and business activity measure or county-wide release amount used to estimate emission rate (lb/yr)
Stationary Area Sources:	
• Household, small office building, and other miscellaneous sources (sources will be combined for modeling)	• County name • Chemical released • CAS number • Emission rate (lb/yr) on a county-wide basis
Mobile Sources:	
• On-road: All trucks, buses, cars, and any other street or highway vehicle • Non-road: Trains, airplanes, ships, construction equipment, lawn equipment	• County name • Chemical released • CAS number • Emission rate (lb/yr) on a county-wide basis
Background Sources:	
• Releases that are not the result of current human activity, including both natural and past human sources	• Chemical name • CAS number • National background concentration measurement ($\mu g/m^3$)
Monitored Concentrations:	• Chemical name • Annual average concentration ($\mu g/m^3$)

What are the steps the Partnership will need to complete to set up the Emission Source Inventory and collect the information needed for the Initial Screen?

The following list of tasks will need to be completed by the Partnership to set up the emission source inventory prior to the Initial Screen:

Step 1: Set up quality assurance/quality control (QA/QC) procedures for data collection and entry

Step 2: Set up an Emission Source Inventory database to incorporate information needed for the screening process

Step 3: Identify and collect location information for all stationary point sources in the study area

Step 4: Collect emission rate information for stationary point sources when release information is available for each individual source

Step 5: Estimate emission rates for stationary point sources when release information is not available for each individual source

Step 6: Identify and collect emission information for stationary area sources (sources combined for modeling)

Step 7: Collect information on mobile source emissions

Step 8: Collect information on background concentrations

Step 9: Collect information on relevant monitored concentrations

How will the Partnership carry out these steps?

Step 1:   Guidance for setting up QA/QC procedures for data collection and entry

Quality assurance and quality control are important considerations for all work performed during the air screening process. The overall quality control objective is to collect well-documented data with known quality. This objective involves establishing and meeting goals for precision, completeness, and representativeness. The conclusions and recommendations made by the Partnership may potentially be used to justify changes for emission sources in the study area. Therefore, the information gathered and used to create the source inventory database must be accurate, complete, and defensible. The Partnership is responsible for reviewing the data collected by the Partnership on each facility in the study area and for deciding whether the data are of appropriate quality for inclusion in the source inventory database.

To ensure the accuracy of the work when building the source inventory database, all information gathered by the Partnership about the facilities should be documented. By documenting the information collected from the emitting facilities, the Partnership will be able to identify where and when a particular value was obtained (e.g., facility records, permits, and telephone conversations). Forms should be created to document telephone conversation, facility visits, and other data collection efforts. Notes and other hard copies of data obtained from the facilities should be filed in a manner that allows for easy retrieval for future reference.

Data collected from the facilities should be recorded on predesigned data entry forms that are printed out and filled in by hand or electronically. The forms should be designed to ensure that a consistent set of information is collected from each emission source during the Initial, Secondary, and Final Screening steps. This minimizes the potential for data gaps and helps to ensure that all needed information about a given emission source is obtained.

The Emission Source Inventory is used to store the data collected by the Partnership. This information will be used throughout the screening process. Therefore, it is important that the information be as accurate and comprehensive as possible. The following QA/QC procedures should be put into place by the Partnership to help ensure the integrity of the database.

Quality control for data entry is an important consideration. Data on air emissions collected by the Partnership should be entered accurately into the database. There is the potential for data entry errors to occur when transcribing information from notes, facility records, and permitting forms, and when importing data from existing databases. Therefore, procedures should be put into place to compare the data entered into the emission source inventory with the original source of the data. This can be accomplished by comparing the entries in the database with the information recorded on the data entry forms. A more stringent quality control procedure involves entering the data into two separate databases and comparing the fields in each database to see if there are discrepancies. Differences between the two databases can then be examined, a decision made as to which values are correct, and appropriate changes made.

Quality control for mathematical manipulations includes checking the calculations used to convert emission rates to ambient air concentrations. In addition, it involves checks to ensure that consistent units are used when comparing ambient air concentration values to screening-level concentration values. Emissions data are provided in different units (e.g., tons, pounds, grams) and for a variety of time periods (e.g., per year, day, hour), and the emissions data may be reported as a rate of emission (e.g., tons/year, pounds/year, grams/hour). The Partnership should be sure to work with consistent units of measure. The mathematics used to convert from one unit of measure to another should be checked for accuracy.

As a part of its quality control, the Partnership will need to pay special attention that releases from sources are not entered into the inventory database more than once. Since emissions data are collected from a variety of sources, there is the potential for information to be collected about the same emission source from more than one source. If the same releases are recorded in more than one database, the Partnership's Inventory Team should only use the information from the database with the best quality and most appropriate data for screening. It is also possible that some releases may have been entered more than once in the same database by error. If there are multiple releases of the same chemical from the same facility, the Inventory Team may want to check to be sure that they are actually different releases and not multiple entries for the same release. Checking to make sure that the release amounts entered into the

Partnership's database are not the result of counting the same release more than once will be a key part of the Partnership's quality control work.

Integrity and defensibility of the database are very important because the information in the Emission Source Inventory will be used by the Partnership as a basis for making decisions about air quality in the study area. The Partnership should provide the opportunity for independent review of the information in the Emission Source Inventory. For example, facilities in the study area should be given the opportunity to review and comment on the accuracy of the data. Data reports that summarize information contained in the source inventory database can also be generated electronically and can be checked for accuracy by the facilities.

A particular chemical may be known by a variety of names (e.g., methylene chloride is also called dichloromethane). For consistency, it is best to use the Chemical Abstracts Service (CAS) registry number in the Emission Source Inventory. The CAS number is provided in the format xxxxx-xx-x. The number is unique for each chemical and allows efficient searching on computerized databases. The CAS number entries should be checked to ensure that the correct number is used.

Step 2: Guidance for setting up the Emission Source Inventory database to incorporate information needed for the screening process

The Emission Source Inventory database will be set up to store the information shown on Table 9-1 in preparation for conducting the Initial Screen. This information is collected by the Partnership for the purpose of identifying as many sources of air emissions as possible (i.e., stationary point sources, stationary area sources, mobile sources, and background and relevant monitored concentrations) within the study area. The database will subsequently be updated during the Initial, Secondary, and Final Screens to include more detailed and accurate information about the chemical emissions that affect the study area.

Figure 9-2 shows an example template for a spreadsheet that can be used to begin developing the emission source inventory. Column A is the Emission Source Type (i.e., stationary point, stationary area, background concentrations, and relevant monitored concentrations). Column B is the Facility or County Name and is used to identify the source of the emission. Column C is the

Chemical Name and is used to identify the chemical that is being released. Column D is the Chemical Abstract Service (CAS) registry number (CAS Number). Column E is the Annual Emission Rate of the chemical from the source in units of pounds per year. Emission rate information is collected at the time for stationary point sources, stationary area sources, and mobile sources. Column F is the Measured Ambient Air Concentration (i.e., monitored concentrations) in units of $\mu g/m^3$. Column G is the Background Source Concentration in units of $\mu g/m^3$.

Step 3: Guidance for identifying and collecting location information for all stationary point sources in the study area

As the first step in collecting the information needed for the screening process, the technical team will need to identify all the stationary point sources of air toxics in the Partnership study area. As discussed in the Overview, the boundaries of the area to be studied will be determined by the Partnership. In many cases, use of ZIP codes to define the boundaries of the study area will make it easier to obtain applicable records for the emission sources contained in the existing databases.

The technical team will not need to collect location information for the stationary area sources in the Partnership area. These sources, such as household heating or consumer product use, are too numerous or irregular to locate individually. The following is a list of the stationary sources that will be treated as area sources and will not require location information. Any stationary source in the Partnership area that does not belong to one of these area source categories will need to be identified, located, and handled as a point source.

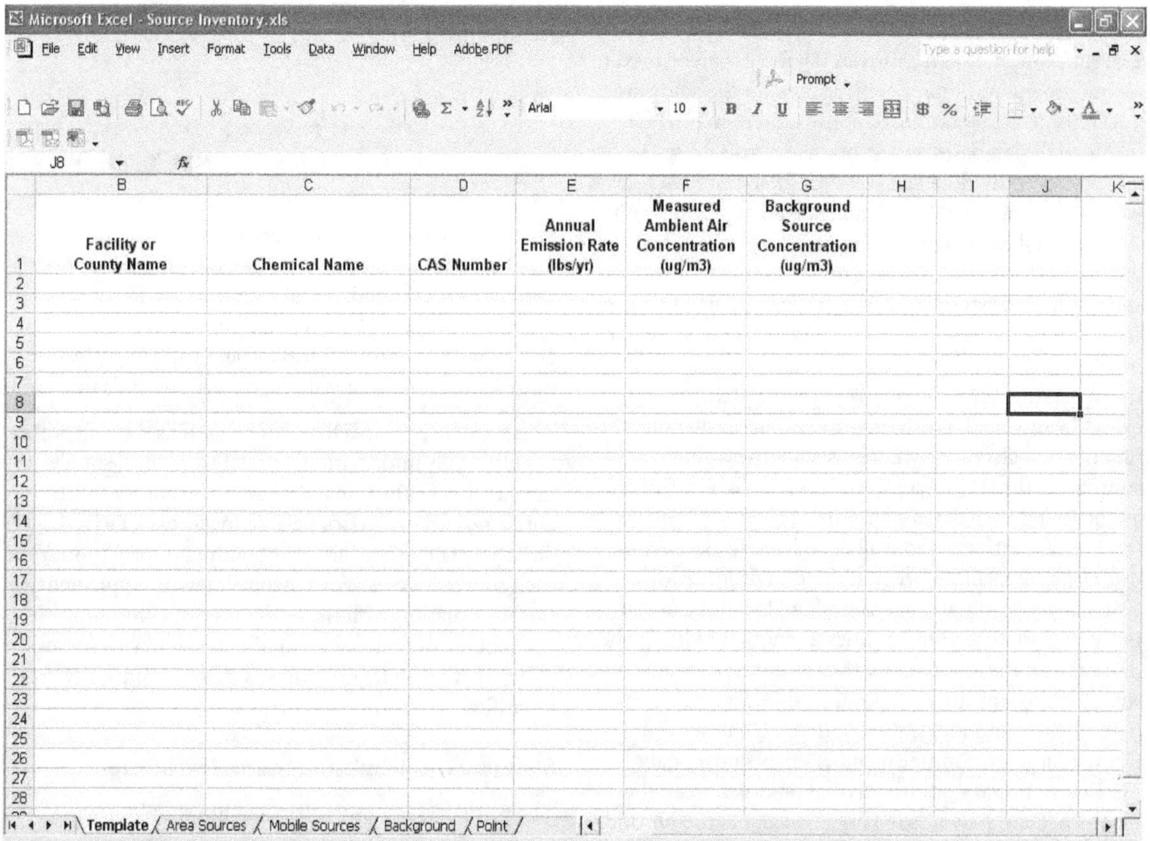

Figure 9-2.
Template for Emission Source Inventory Database

Stationary sources that will be handled as area sources include:

- Agricultural production
- Asphalt paving: Cutback asphalt
- Consumer products usage
- Gasoline distribution Stage 1
- Industrial boilers: Distillate oil
- Institutional/commercial heating (all types)
- Natural gas transmissions and storage
- Non-industrial asphalt roofing
- Pesticide application
- Residential heating (all types)
- Residential heat (wood)
- Structure fires
- Surface coatings: Architectural
- Surface coatings: Industrial maintenance
- Surface coatings: Traffic markings

The state and national air toxics databases can be used to identify many stationary point sources in the study area, but in some areas stationary point sources may be included in databases as part of an area source, so the location of individual facilities and their emission rates will not be available. The technical team, working with the Partnership, will need to examine the available databases to determine which of the stationary point sources in their study area are handled as part of an area source or were missed as point sources. These stationary point sources need to be identified and located to ensure that all stationary point sources are included. For example, the databases covering some areas do not identify dry-cleaning facilities as stationary point sources. In these databases, releases from dry-cleaners are included in the area source database as a county-wide total release estimate from all dry-cleaners. Because the dry-cleaning facilities are combined to produce a county-wide estimate, the names and locations of the individual dry-cleaning facilities are not included in the database. In these situations, the technical team will need to rely on community and local business members to help identify and locate these sources. If the study area is not too large, committee members may know of all existing stationary point sources. For larger areas, the technical team may need to organize an on-the-ground survey to find and identify all sources. Available business databases may also be used to identify any business sources that may not be included in the State or national

air toxics database. For example, the Dunn and Bradstreet database lists all businesses by ZIP code, SIC code, and address, and this information can be used to help identify potential point sources not identified in the state and local databases. By combining all the approaches described above and relying on the work and knowledge of all the members of the Partnership, the technical team will be able to identify all stationary point sources of air toxics in the study area. Collecting or estimating release information for these sources will be discussed in the next steps.

To ensure that all sources are identified and located properly, the Partnership may find it useful to develop a map, either using GIS or hard copy, to show all sources and their locations. This will help the community and business members of the Partnership identify the sources and add ones that may be missing. A map locating all stationary point sources, major roadways, and major non-road mobile sources will also be useful as an educational tool for meetings with the broader community.

Step 4: Guidance for collecting release information for stationary point source emissions when release information is available in existing databases for each individual source

Figure 9-3 shows the procedure for adding stationary point source emissions data when release information is available for each individual source to the Emission Source Inventory. The information required is the facility name, facility location, chemical name, CAS number, and emission rate (in units of pounds per year).

There are three places that can be used as starting points to collect this information: (1) state and local source inventories, (2) National Emissions Inventory (NEI) database, and (3) Toxics Release Inventory (TRI) database. The procedure for obtaining information from each source follows. The Partnership will supplement this information with the knowledge of community residents and businesses and, if necessary, with surveys, to ensure that all facility sources in the community are identified.

State and Local Emission Source Inventories

The primary sources the Partnership should use for information about stationary point source emissions data are state and local inventories. State and local government authorities are responsible for permitting air releases, and they maintain databases containing information about local sources and their releases. These databases and associated files contain information on the chemicals regulated by state and local legislation. The amount and kind of information collected will vary depending on the requirements of the local authority. In some cases, the information may be obtained from the Internet. For example, a listing of selected state and local agencies providing emissions data can be found on the Internet at http://www.epa.gov/ttn/chief/related.html#state.

Personnel from local government staff, members of environmental organizations, and university staff will be familiar with the contents of state and local databases. A representative from the state who is already familiar with the state and local databases will likely be a member of the Partnership and the Emission Source Inventory Technical Team. This person would be a good choice for taking the lead in acquiring information from the state and local databases. Staff in the air permitting office in the state or local government will be familiar with sources of emissions data within the study area and how they can be accessed.

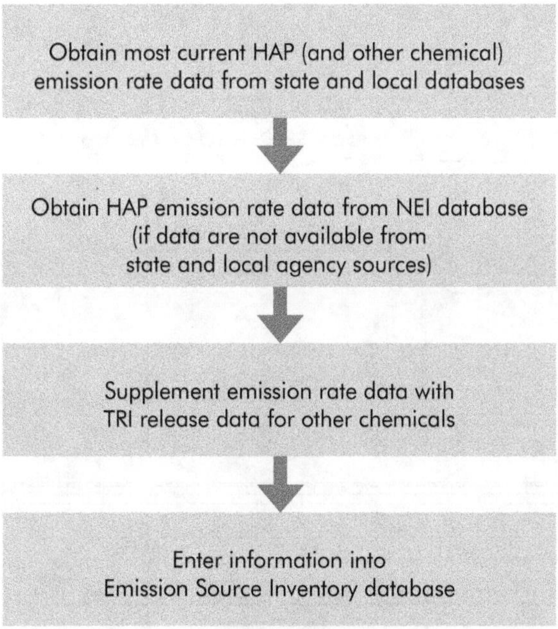

National Emissions Inventory (NEI) Database

Information about the NEI database and instructions for downloading the data files for each state are provided at http://www.epa.gov//ttn/chief/net/1999inventory.html.

The Zip file for the NEI point source files decompress to a single data file with the naming convention xx99ptfinal.mdb. For example, the Maryland data file would be called MD99ptfinal.mdb. The file contains a total of nine data tables that can be viewed and manipulated using Microsoft Access. The nine tables are identified as follows:

1. tblPointCE
2. tblPointEM
3. tblPointEP
4. tblPointER
5. tblPointEU
6. tblPointPE
7. tblPointSI
8. tblPointTR
9. tblRecordCount

Only selected information from these tables is needed to begin development of the Emission Source Inventory. The required NEI information for stationary point sources, along with the corresponding data elements to be extracted from the point source data tables listed above, are provided in Table 9-2. Microsoft Access is used to extract the data from the NEI point source data files by setting up the query design view screen as shown in Figure 9-4. Initially only three of the eight data tables (i.e., tblPointSI, tblPointEM, and tblPointER) are needed. Figure 9-4 shows how to relate these tables and shows the structure to be used for the query. Information is obtained by searching for release rate information about the facilities located within a specific county. This is accomplished by selecting records containing the county name or the names of the facilities located within the study area using the field tblPointSI –> strFacilityName. Once the query has been run, the resulting information shown on Table 9-2 is imported into the Emission Source Inventory. To obtain the estimated annual emission rate, it is necessary to combine data from three separate fields in the tblPointEM table (i.e., dblEmissionNumericValue, strEmissionUnitNumerator, strEmissionType) and, in some cases, to convert the units for the resulting emission rate to pounds per year. The field strEmissionType in the tblPointEM table refers to

**Figure 9-3.
Procedure for Adding Stationary Point Source Emissions Data to the Emissions Source Inventory (when release information IS available for each individual source)**

emission type that was reported. The emission types include actual annual, average daily, average hourly, maximum allowable annual, maximum annual, maximum daily, maximum hourly, and potential annual emissions. Figure 9-5 shows an example of the Emission Source Inventory database with sample stationary point source data entered.

Toxics Release Inventory (TRI) Database

The Toxics Release Inventory (TRI) is a publicly available EPA database that contains information on toxic chemical releases and other waste management activities reported annually by certain covered industry groups as well as federal facilities. TRI contains information about toxic chemicals used, manufactured, stored, treated, transported, or released into the environment. Information about TRI is available on the EPA Internet site at http://www.epa.gov/tri.

TRI can be used to supplement information on HAP

emissions that were obtained from the state and local source inventories and from NEI. TRI contains data on the releases of more than 600 designated toxic chemicals to air, water, and land. The list of chemicals can be downloaded as a PDF file from http://www.epa.gov//tri/chemical/hemlist2001.pdf.

TRI lists releases for more than 600 chemicals, while NEI lists the releases for only 188 hazardous air pollutants (HAPs) (shown in Appendix E). However, because facilities must exceed a certain emission threshold before reporting releases, certain releases and facilities may not be available in TRI that may be contained in other resources. TRI also is only applicable for facilities that fall under the predefined Standard Industrial Classification (SIC) codes. The SIC codes that are covered by TRI are available at http://www.epa.gov/triinter/report/siccode.htm.

TRI data can be obtained from several sources described

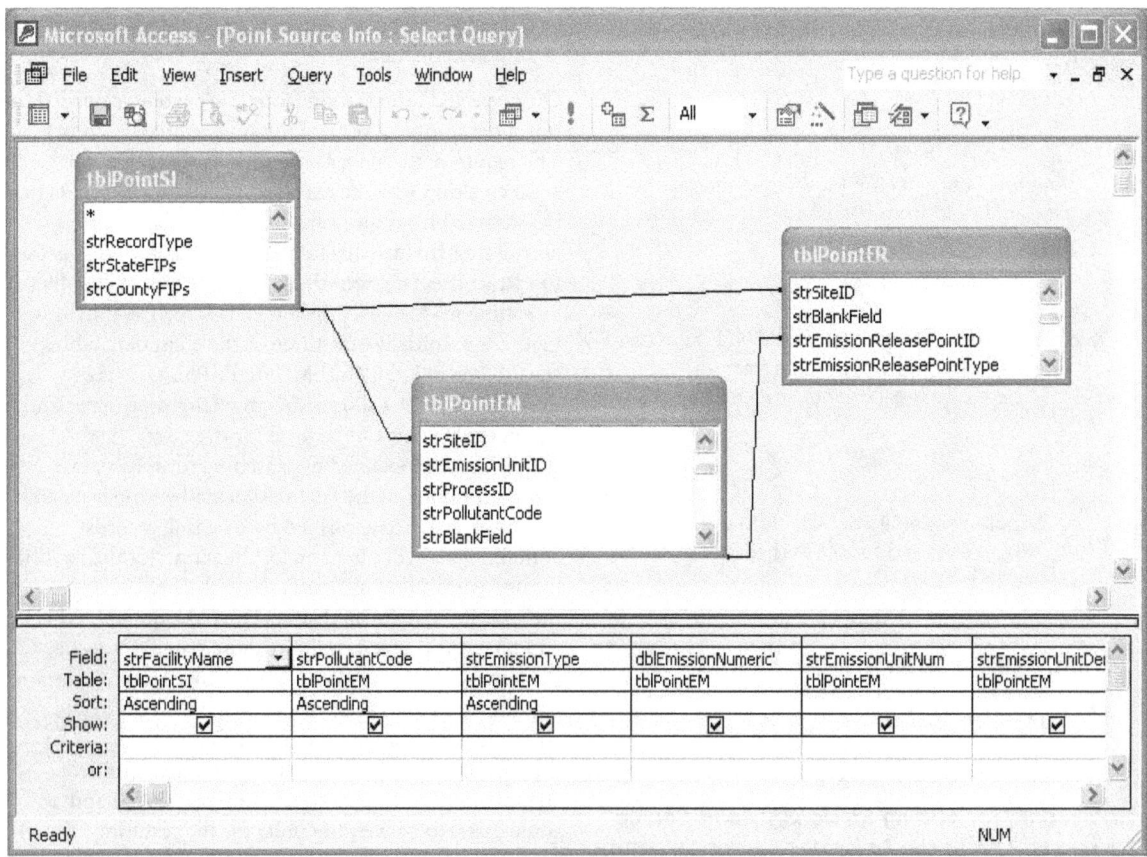

Figure 9-4.
Microsoft Access Query Design View Screen for NEI Point Sources

at http://www.epa.gov/triinter/tridata/index.htm.

TRI Explorer provides fast and easy access to the TRI data to identify facilities and chemical releases that warrant further study and analysis. The site provides user-definable reports to obtain information on releases.

Public Data Releases are available at http://www.epa.gov/triinter/tridata/index.htm#pdr.

The annual TRI Public Data Release includes a general overview of that year's TRI data and information on trends, state fact sheets that provide a brief summary of the TRI data by state, and downloadable data files containing TRI reports submitted for the reporting year.

TRI State Data Files and documentation describing the contents of the files are available at http://www.epa.gov/triinter/tridata//state_data_files.htm.

The TRI State Data Files are sets of files containing all data submitted to the Toxic Chemical Release Inventory by facilities located in a selected state for a specific year. The data has been extracted from the Toxics Release Inventory System (TRIS).

The state files are downloaded as executable files with the format xx.exe, where xx is the state abbreviation (e.g., the Maryland file would be called MD.exe). Three text files are created after executing the xx.exe file: (1) xx_St1.TXT, (2) xx_St2.TXT, and (3) xx_St3.TXT. Only selected information from file xx_St1.TXT is needed to begin development of the Emission Source Inventory. The required information about stationary point sources to be included in the Emission Source Inventory, along with the corresponding data elements to be extracted from the TRI xx_St1.TXT point source file, are provided in Table 9-3.

Step 5: *Guidance for estimating emission rates for stationary point sources when release information is __not__ available for each individual source*

Sometimes release information for a stationary point source cannot be found in state and local databases. There are two ways in which this could happen. Either a facility located in the study area may, for some reason, not be included in the stationary point source database even though other facilities of its type are included, or the databases handle facilities of this type as an area source, and information on releases is available, but only as a county-wide total for all facilities of the same type and not for each facility. Since emissions rates for these individual stationary point source facilities are not available in either of these cases, the technical team will need to estimate these releases.

The most accurate way to estimate emissions for a facility uses emission factors and source-specific use

Table 9-2.
Required Data Elements for Emission Source Inventory from NEI Point Source Files

EMISSION SOURCE INVENTORY DATABASE FIELD	TABLE AND FIELD NAME FROM NEI POINT SOURCE DATA FILE
Facility or County Name	tblPointSI –> strFacilityName
Facility Location	tblPointSI –> strLocationAddress tblPointSI –> strCity tblPointSI –> strState tblPointSI –> strZipCode
Chemical Name	The NEI point source files contain information on the CAS number but do not have a field for the chemical name. The Partnership will need to look up the chemical name associated with the CAS number and enter that information.
CAS Number	tblPointEM –> strPollutantCode
Estimated Annual Emission Rate (pounds/year)	tblPointEM –> dblEmissionNumericValue tblPointEM –> strEmissionUnitNumerator tblPointEM –> strEmissionType

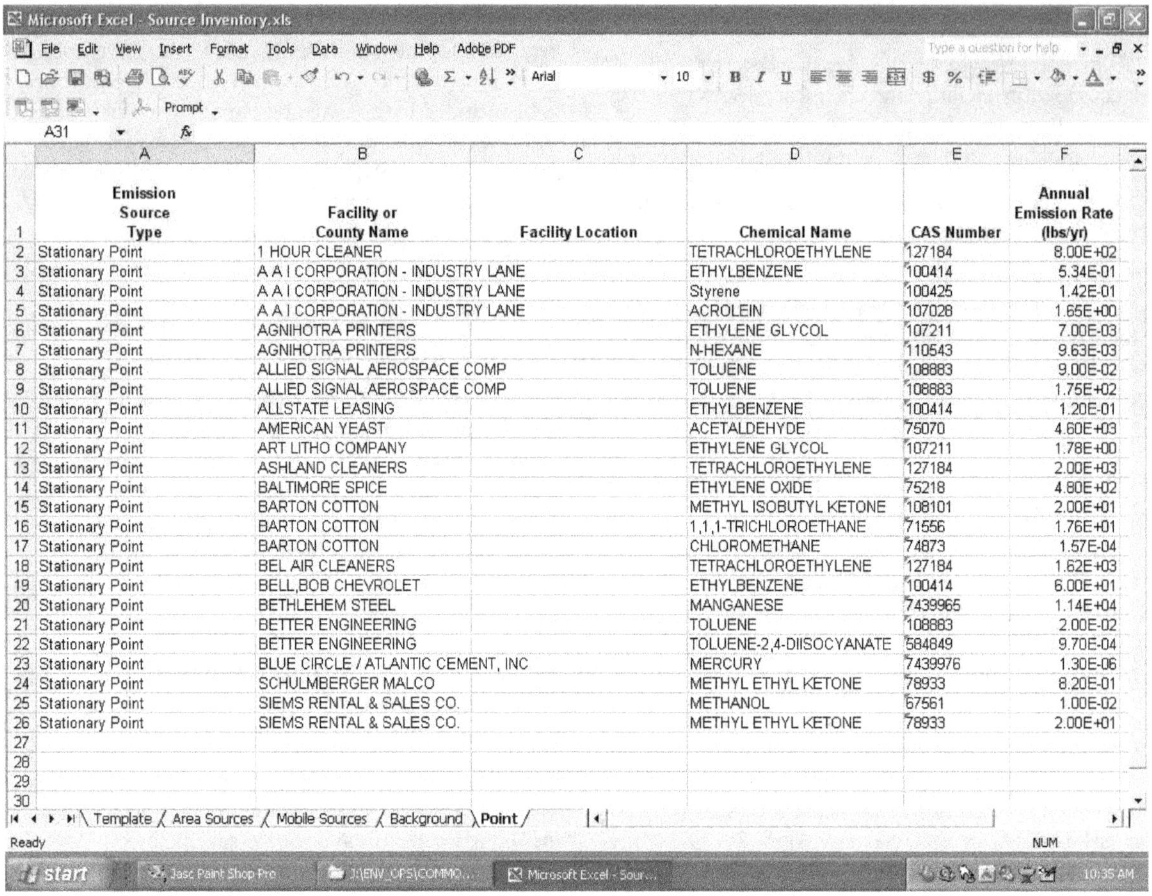

	Emission Source Type	Facility or County Name	Facility Location	Chemical Name	CAS Number	Annual Emission Rate (lbs/yr)
2	Stationary Point	1 HOUR CLEANER		TETRACHLOROETHYLENE	127184	8.00E+02
3	Stationary Point	A A I CORPORATION - INDUSTRY LANE		ETHYLBENZENE	100414	5.34E-01
4	Stationary Point	A A I CORPORATION - INDUSTRY LANE		Styrene	100425	1.42E-01
5	Stationary Point	A A I CORPORATION - INDUSTRY LANE		ACROLEIN	107028	1.65E+00
6	Stationary Point	AGNIHOTRA PRINTERS		ETHYLENE GLYCOL	107211	7.00E-03
7	Stationary Point	AGNIHOTRA PRINTERS		N-HEXANE	110543	9.63E-03
8	Stationary Point	ALLIED SIGNAL AEROSPACE COMP		TOLUENE	108883	9.00E-02
9	Stationary Point	ALLIED SIGNAL AEROSPACE COMP		TOLUENE	108883	1.75E+02
10	Stationary Point	ALLSTATE LEASING		ETHYLBENZENE	100414	1.20E-01
11	Stationary Point	AMERICAN YEAST		ACETALDEHYDE	75070	4.60E+03
12	Stationary Point	ART LITHO COMPANY		ETHYLENE GLYCOL	107211	1.78E+00
13	Stationary Point	ASHLAND CLEANERS		TETRACHLOROETHYLENE	127184	2.00E+03
14	Stationary Point	BALTIMORE SPICE		ETHYLENE OXIDE	75218	4.80E+02
15	Stationary Point	BARTON COTTON		METHYL ISOBUTYL KETONE	108101	2.00E+01
16	Stationary Point	BARTON COTTON		1,1,1-TRICHLOROETHANE	71556	1.76E+01
17	Stationary Point	BARTON COTTON		CHLOROMETHANE	74873	1.57E-04
18	Stationary Point	BEL AIR CLEANERS		TETRACHLOROETHYLENE	127184	1.62E+03
19	Stationary Point	BELL,BOB CHEVROLET		ETHYLBENZENE	100414	6.00E+01
20	Stationary Point	BETHLEHEM STEEL		MANGANESE	7439965	1.14E+04
21	Stationary Point	BETTER ENGINEERING		TOLUENE	108883	2.00E-02
22	Stationary Point	BETTER ENGINEERING		TOLUENE-2,4-DIISOCYANATE	584849	9.70E-04
23	Stationary Point	BLUE CIRCLE / ATLANTIC CEMENT, INC		MERCURY	7439976	1.30E-06
24	Stationary Point	SCHULMBERGER MALCO		METHYL ETHYL KETONE	78933	8.20E-01
25	Stationary Point	SIEMS RENTAL & SALES CO.		METHANOL	67561	1.00E-02
26	Stationary Point	SIEMS RENTAL & SALES CO.		METHYL ETHYL KETONE	78933	2.00E+01

Figure 9-5.
Sample Stationary Point Source Data in Emission Source Inventory Database

Table 9-3.
Required Data Elements for Emission Source Inventory from TRI Point Source Files

EMISSION SOURCE INVENTORY DATABASE FIELD	FILE AND FIELD NAME FROM TRI POINT SOURCE DATA FILE
Facility or County Name	xx_St1.txt –> Facility Name
Facility Location	xx_St1.txt –> Facility Street xx_St1.txt –> Facility City xx_St1.txt –> Facility State xx_St1.txt –> Facility ZIP Code
Chemical Name	xx_St1.txt –> Chemical Name
CAS Number	xx_St1.txt –> CAS Number
Estimated Annual Emission Rate (pounds/year)	xx_St1.txt –> Total Stack Air Emissions or xx_St1.txt –> Total Fugitive Air Emissions

Note: The TRI database reports the emission rate in units of pounds per year.

information (i.e., bottom-up approach). This bottom-up approach is recommended if sufficient resources are available to collect source-specific information. As a less resource-intensive alternative to the bottom-up approach, information about individual stationary point source emissions may be developed using a surrogate, such as percent of population, to apportion the county-wide emissions to each facility (i.e., top-down approach). Examples of procedures for developing emission estimates using the top-down and bottom-up approaches are discussed below.

Dry-Cleaner Example

The Partnership determines that there are four dry-cleaners (i.e., dry-cleaner A, B, C, and D) located in the study area. Information is not available about emissions from each of the individual dry-cleaners, but an aggregate county-wide estimate for the emissions from all the dry-cleaners in the county is available. The Partnership can proceed with developing emissions on a per facility basis using either a top-down or bottom-up approach. Figure 9-6 shows the procedure for adding stationary point source emissions data to the Emission

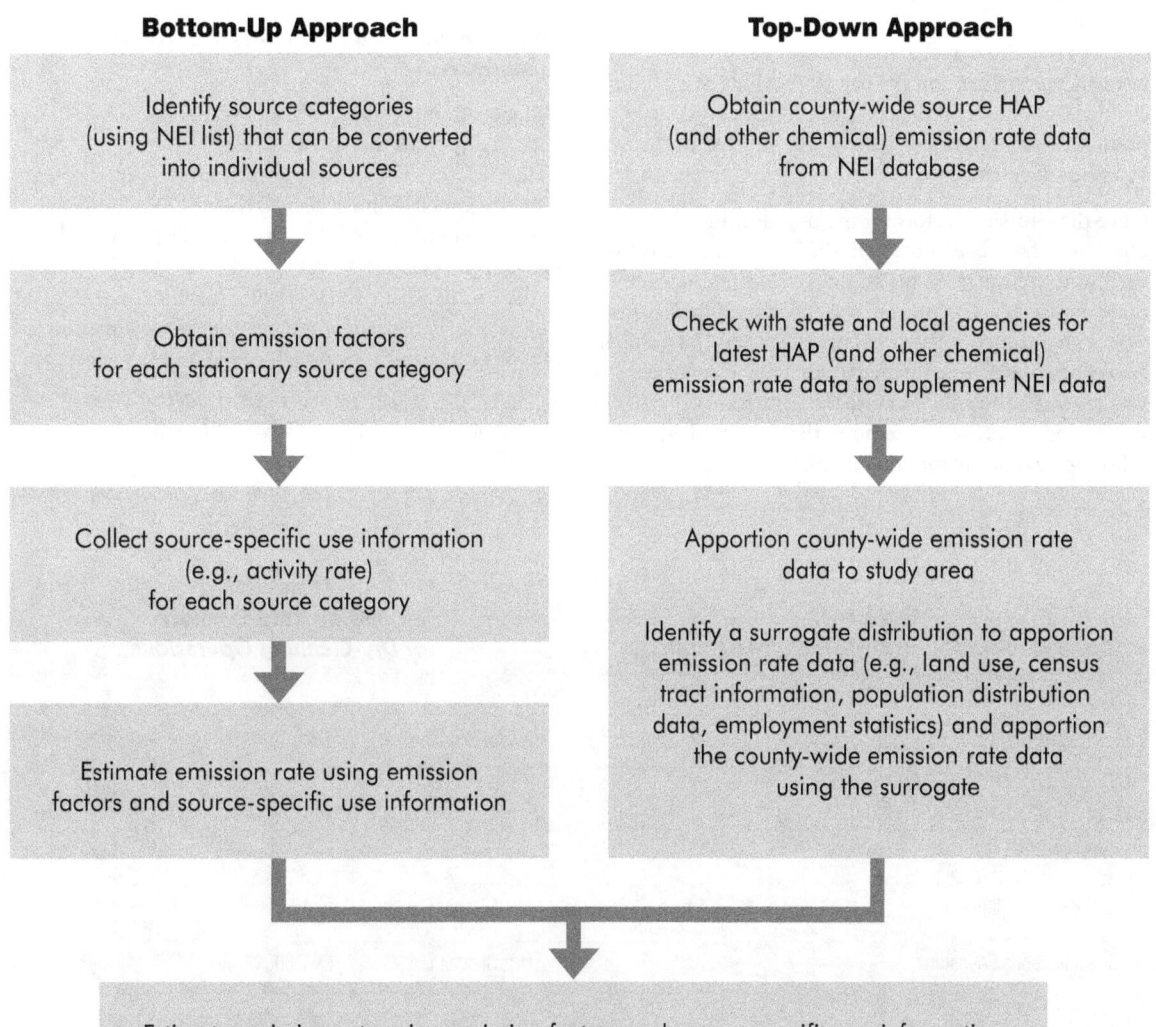

Figure 9-6.
***Procedure for Adding Stationary Point Source Emissions Data to Emission Source Inventory
(when release information IS NOT available for each individual source)***

Source Inventory, using the bottom-up and top-down approaches, when release information is not available for each individual source.

Bottom-Up Approach

The Compilation of Air Pollutant Emission Factors (AP-42, Vol. 1, 5th ed. Stationary and Area Sources, January, 1995) provides emission factors for activities that generate air emissions. The compilation of emission factors can be found on-line at http://www.epa.gov/ttn/chief/ap42/.

Table 9-4 shows emission factors from AP-42 for dry-cleaning operations. The emission factor for dry-cleaners is defined as the number of pounds of tetrachloroethylene emitted per 100 pounds of clothing cleaned. As shown on the table, approximately 27.5 pounds of tetrachloroethylene will be emitted to clean 100 pounds of clothing.

Once the emission factors for the dry-cleaning operations have been obtained, the Partnership calls the various dry-cleaners in the study area to find out how many pounds of clothing are washed at each facility per year. The product of the emission factor and the number of pounds of clothing cleaned at each dry-cleaner per year provides the emission rate for tetrachloroethylene in pounds per year. Table 9-5 shows the estimated emission rates for each of the four dry-cleaners.

Top-Down Approach

Using the procedure described in step 5 for obtaining information from the NEI area source database about tetrachloroethylene dry-cleaners, the Partnership determines that 41.04 tons of tetrachloroethylene are released each year in the county that encompasses the study area. Using a surrogate, the emission rate for the county is apportioned to the study area. There are many surrogates that the Partnership can use to apportion emissions data using the top-down approach. Two examples of surrogates that can be used for the top-down approach for this example are provided below.

Apportioning Based on Number of Facilities (county vs. study area)

Using economic census data as a surrogate, the Partnership finds that 107 businesses fall under NAICS code 81232 (i.e., dry-cleaning and laundry services) and 35 businesses fall under NAICS code 81231 (i.e., coin-operated laundries and dry-cleaners). The total of these two categories is 107 + 35 = 142 dry-cleaning businesses in the county. Using the following calculation, the Partnership can apportion the county-wide emission rate of 41.04 tons/year to each of the four dry-cleaners:

$$\frac{41.04 \text{ tons/yr} \times 2,000 \text{ lb/ton}}{142 \text{ facilities in county}} = \begin{array}{l}578.03 \text{ lb/year tetrachloroethylene} \\ \text{released per dry-cleaner}\end{array}$$

Table 9-4.
Emission Factors for Dry-Cleaning Operations

OPERATION	EMISSION FACTOR* (POUNDS/100 LB)
Washer/Dryer/Still/Muck Cooker	8
Filter Disposal	16.4
Still Residue Disposal	1.6
Miscellaneous Emissions	1.5
Total	27.5

*The emission factor is defined as the pounds of tetrachloroethylene per 100 pounds of clothes cleaned.
Source: Compilation of Air Pollutant Emission Factors, AP-42, Vol. 1, 5th ed. Stationary and Area Sources, January 1995.

Table 9-5.
Emission Rate Estimate for Dry-Cleaning Operations

DRY-CLEANER	CLOTHES TREATED/YR (POUNDS)	EMISSION RATE (POUNDS/YR)
Dry-Cleaner A	2,000	550
Dry-Cleaner B	2,400	660
Dry-Cleaner C	1,600	440
Dry-Cleaner D	1,300	357.5

Note: The emission rate is calculated by multiplying the total emission factor (i.e., 27.5 lb/100 lb) from Table 9-4 by the pounds of clothes treated per year for each dry-cleaner.

Apportioning Based on Population (county vs. study area)

Using census tract data as a surrogate, the Partnership finds that the total population in the county is 692,134. The census tract data also show that the population in the study area is 17,012. These data show that 17,012/692,134 (i.e., 2.46%) of the population in the county are in the study area. Using this percentage as a surrogate, the Partnership can apportion the county-wide emission rate of 41.04 tons/year to each of the four dry-cleaners using the following calculation:

$$\frac{41.04 \text{ tons/yr} \times 2{,}000 \text{ lb/ton} \times (2.46/100)}{\text{dry cleaner}} = \frac{504.79 \text{ lb/year}}{\text{tetrachloroethylene released/dry cleaner}}$$

Step 6: *Guidance for collecting information needed for stationary area source emissions (sources that will be combined for modeling)*

Figure 9-7 shows the procedure for adding stationary area source emissions data to the Emission Source Inventory. Stationary area source emissions are defined as county-wide emission rates that cannot be converted into individual point sources using one of the methods described in step 4. These sources cannot be turned into point sources because they are too numerous or irregular for point source modeling. The technical team will find and enter the county-wide emissions total for each area source category into the Emission Source Inventory database.

Emissions Data for Emission Source Inventory

As discussed in step 3 above, using the categories established in the National Toxics Inventory, the following will be treated as stationary area sources:

- Agricultural production
- Asphalt paving: Cutback asphalt
- Consumer products usage
- Gasoline distribution Stage 1
- Industrial boilers: Distillate oil
- Institutional/commercial heating (all types)
- Natural gas transmissions and storage
- Non-industrial asphalt roofing
- Pesticide application
- Residential heating (all types)
- Residential heat (wood)
- Structure fires
- Surface coatings: Architectural
- Surface coatings: Industrial maintenance
- Surface coatings: Traffic markings

Obtain county-wide source HAP (and other chemical) emission rate data from NEI database (refer to list of source categories that are recommended to be modeled as combined area source)

Check with state and local agencies for latest HAP (and other chemical) emission rate data to supplement NEI data

Enter information into Emission Source Inventory database

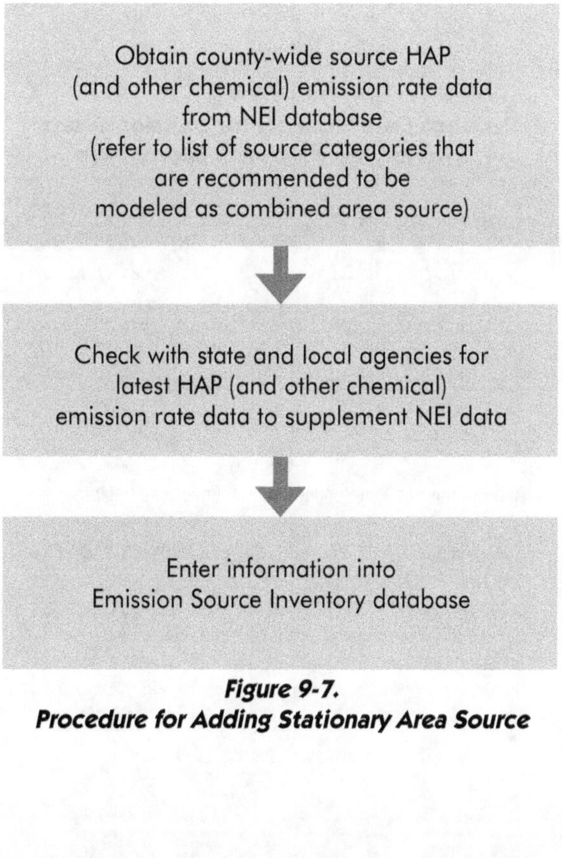

Figure 9-7.
Procedure for Adding Stationary Area Source

Table 9-6.
Data Elements in the 1996 NEI Area Source State Data Files

DATA ELEMENT	DESCRIPTION
Revision Indicator	O = Original from May 1999 A = New record added by state/trade association RA = Revised record added by state/trade association
Inventory Year	Baseline year for inventory
State FIPS	2-digit Federal Information Processing Standards state code
County FIPS	3-digit FIPS county code
County Name	Name that corresponds to 3-digit FIPS county code
Area Source Category Name	NEI source category name
SIC Code	Source category Standard Industrial Classification code (when applicable)
EPA MACTID	Source category Maximum Achievable Control Technology code (when applicable)
Pollutant CAS	Unique pollutant Chemical Abstracts Service number
Pollutant Name common name	Hazardous air pollutant
Emissions	Emissions estimate at county level
Emission Units	Units for county emissions estimate
SCC	Source category Source Classification Code
AMS	Source category Area and Mobile Source code
Source Type	Source classification (e.g., area source)

The primary source of stationary area source emissions data is the NEI. The information obtained from NEI will be supplemented with information on HAPs and other chemical emissions obtained from state and local inventories. The area source files consist of a single standard ASCII comma-delimited format file for each state that can be imported into database or spreadsheet programs. A listing of the data elements contained in the state data files are provided on Table 9-6. The Area Source Category Name field contains the name of the area source category.

The required information about stationary area sources to be included in the Emission Source Inventory, along with the corresponding data elements to be extracted from the NEI area source data file, are provided in Table 9-7. Figure 9-8 shows an example of the Emission Source Inventory with sample stationary area source data entered.

The Partnership should check to make sure that all the sources listed in the NEI area source data file are included in the source inventory, either by being converted to a point source or by being treated as a stationary area source.

Table 9-7.
Required Data Elements for Emission Source Inventory from NEI Area Source Files

EMISSION SOURCE INVENTORY DATABASE FIELD	FIELD NAME FROM NEI AREA SOURCE DATA FILE
Facility or County Name	County name
Name of Chemical Released	Pollutant name
CAS Number	Pollutant CAS
Annual Emission Rate (pounds/year)	Emissions (refer to note)

Note: The emission rate provided in the NEI area source data file may need to be converted to units of pounds per year. The value of the emission is provided in the field labeled Emissions, and the unit for the emission is provided in the field labeled Emission Units.

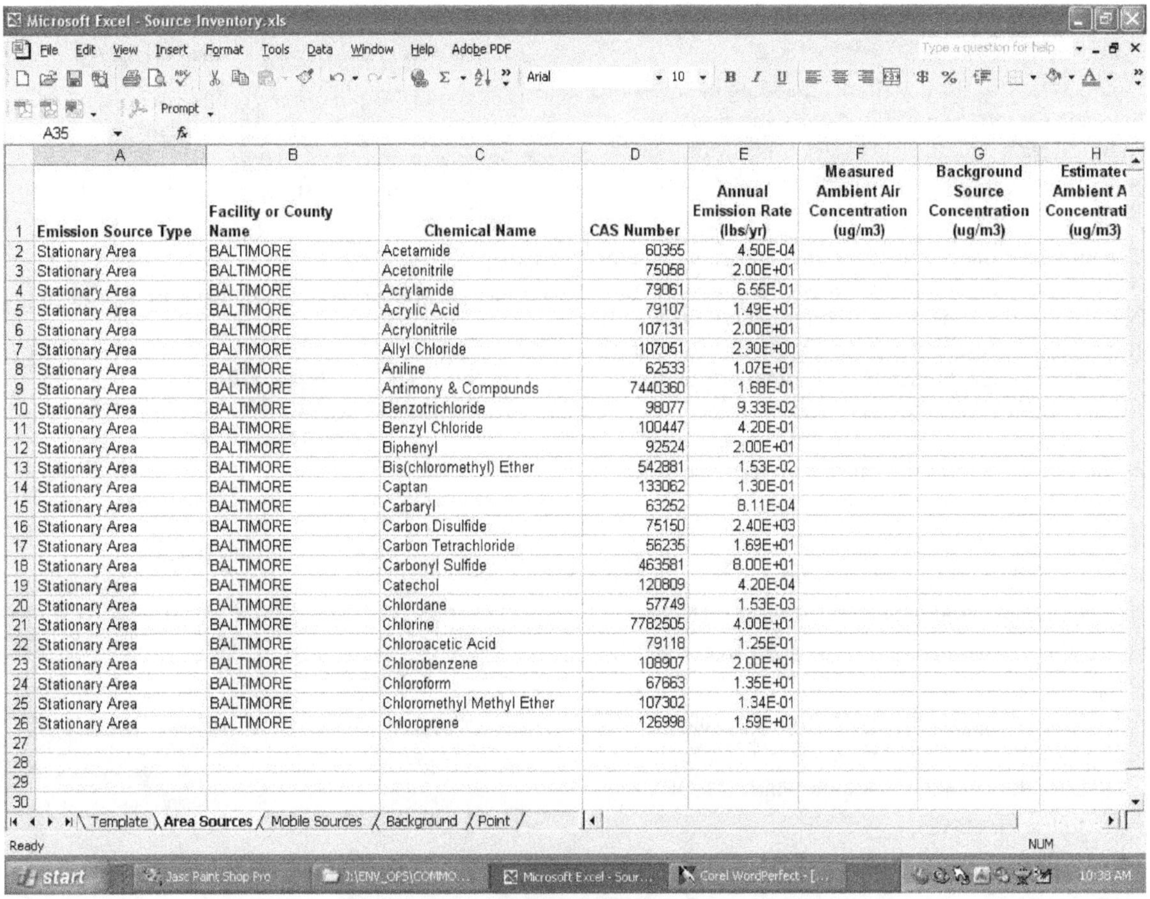

	A	B	C	D	E	F	G	H
1	Emission Source Type	Facility or County Name	Chemical Name	CAS Number	Annual Emission Rate (lbs/yr)	Measured Ambient Air Concentration (ug/m3)	Background Source Concentration (ug/m3)	Estimate Ambient A Concentrati (ug/m3)
2	Stationary Area	BALTIMORE	Acetamide	60355	4.50E-04			
3	Stationary Area	BALTIMORE	Acetonitrile	75058	2.00E+01			
4	Stationary Area	BALTIMORE	Acrylamide	79061	6.55E-01			
5	Stationary Area	BALTIMORE	Acrylic Acid	79107	1.49E+01			
6	Stationary Area	BALTIMORE	Acrylonitrile	107131	2.00E+01			
7	Stationary Area	BALTIMORE	Allyl Chloride	107051	2.30E+00			
8	Stationary Area	BALTIMORE	Aniline	62533	1.07E+01			
9	Stationary Area	BALTIMORE	Antimony & Compounds	7440360	1.68E-01			
10	Stationary Area	BALTIMORE	Benzotrichloride	98077	9.33E-02			
11	Stationary Area	BALTIMORE	Benzyl Chloride	100447	4.20E-01			
12	Stationary Area	BALTIMORE	Biphenyl	92524	2.00E+01			
13	Stationary Area	BALTIMORE	Bis(chloromethyl) Ether	542881	1.53E-02			
14	Stationary Area	BALTIMORE	Captan	133062	1.30E-01			
15	Stationary Area	BALTIMORE	Carbaryl	63252	8.11E-04			
16	Stationary Area	BALTIMORE	Carbon Disulfide	75150	2.40E+03			
17	Stationary Area	BALTIMORE	Carbon Tetrachloride	56235	1.69E+01			
18	Stationary Area	BALTIMORE	Carbonyl Sulfide	463581	8.00E+01			
19	Stationary Area	BALTIMORE	Catechol	120809	4.20E-04			
20	Stationary Area	BALTIMORE	Chlordane	57749	1.53E-03			
21	Stationary Area	BALTIMORE	Chlorine	7782505	4.00E+01			
22	Stationary Area	BALTIMORE	Chloroacetic Acid	79118	1.25E-01			
23	Stationary Area	BALTIMORE	Chlorobenzene	108907	2.00E+01			
24	Stationary Area	BALTIMORE	Chloroform	67663	1.35E+01			
25	Stationary Area	BALTIMORE	Chloromethyl Methyl Ether	107302	1.34E-01			
26	Stationary Area	BALTIMORE	Chloroprene	126998	1.59E+01			
27								
28								
29								
30								

Figure 9-8.
Example Stationary Area Source Data

Step 7: Guidance for collecting information on mobile source emissions

Figure 9-9 shows the procedure for adding mobile source emission data to the Emission Source Inventory. The technical team will find and enter county-wide mobile source emission category totals in the Emission Source Inventory database.

The primary source of mobile source emissions data is NEI. The information on mobile source emissions from NEI will be supplemented with information obtained from state and local inventories. The mobile source files consist of a single standard ASCII comma-delimited format file for each state that can be imported into database or spreadsheet programs. A listing of data elements contained in the state data files are provided in Table 9-8.

The on-road vehicle population is characterized into seven individual vehicle type categories in NEI. They are:

• Light duty gasoline-powered vehicles (LDGV),
• Light duty gasoline-powered trucks up to 6,000 lbs gross vehicle weight (LDGT1),
• Light duty gasoline-powered trucks from 6,000 to 8,500 lbs gross vehicle weight (LDGT2),
• Heavy duty gasoline-powered vehicles (HDGV),
• Motorcycles (MC),
• Light duty diesel-powered vehicles (LDDV), and
• Heavy duty diesel-powered vehicles (HDDV).

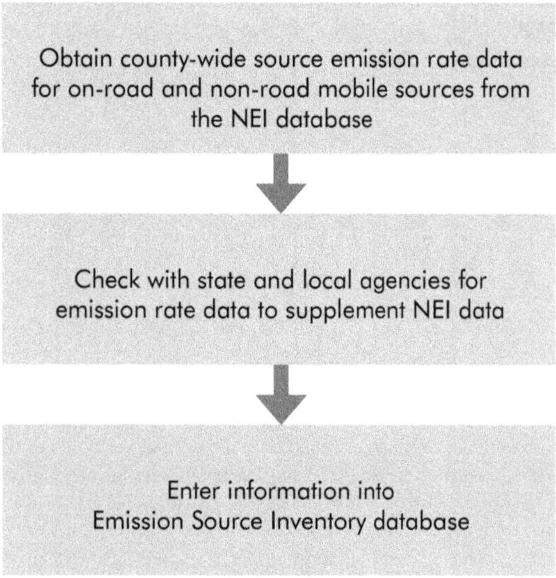

LDGT1 and LDGT2 have been combined into an LDGT category. For some pollutants, the MC category was combined with other vehicle types such as LDGV or HDGV, depending upon the specificity of the data used to estimate the emissions.

The non-road mobile source category in the NEI includes vehicles and equipment that normally are not

Obtain county-wide source emission rate data for on-road and non-road mobile sources from the NEI database

Check with state and local agencies for emission rate data to supplement NEI data

Enter information into Emission Source Inventory database

Figure 9-9.
Procedure for Adding Mobile Source Emissions Data to Emission Source Inventory

Table 9-8.
Data Elements in the 1996 NEI Mobile Source State Data Files

DATA ELEMENT	DESCRIPTION
Revision Indicator	"0" indicates EFIG- or OMS-developed estimates
Start Date	Beginning time for inventory year
End Date	Ending time for inventory year
Country FIPS	FIPS country code
State FIPS	2-digit FIPS state code
County FIPS	3-digit FIPS county code
County Name	Name that corresponds to 3-digit FIPS county code
Mobile Source Category Name	Description of AMS code
Pollutant CAS	Unique NEI pollutant code number (CAS number if available)
Pollutant Name	Hazardous air pollutant common name
Emissions	Emissions estimate at county level
Emission Units	Units for county emissions estimate
Emission	Actual annual air emissions
Source Type	Mobile

operated on public roads to provide transportation. This includes categories such as lawn and garden equipment, agricultural equipment, logging equipment, construction equipment, airport service vehicles, aircraft, locomotives, or commercial marine vessels, and recreational equipment including recreational marine equipment. Gasoline-powered non-road vehicles and equipment can be characterized into two individual engine-type categories, specifically 2- and 4-stroke engines. To estimate the metallic pollutant emissions in this inventory, the 2- and 4-stroke engine-type categories are combined into one category called gasoline engines.

The required information about mobile sources to be included in the Emission Source Inventory, along with the corresponding data elements to be extracted from the NEI mobile source data file, are provided in Table 9-9. Figure 9-10 shows an example of the Emission Source Inventory with sample mobile source data entered.

Table 9-9.
Required Data Elements for Emission Source Inventory from NEI Mobile Source Files

EMISSION SOURCE INVENTORY DATABASE FIELD	FIELD NAME FROM NEI AREA SOURCE DATA FILE
Facility or County Name	County name
Mobile Source Category	Mobile source category name
Name of Chemical Released	Pollutant name
CAS Number or Unique NEI Pollutant Code Number	Pollutant CAS
Annual Emission Rate (pounds/year)	Emissions (refer to note)

Note: The emission rate provided in the NEI mobile source data file may need to be converted to units of pounds per year. The value of the emission is provided in the field labeled Emissions, and the unit for the emission is provided in the field labeled Emission Units.

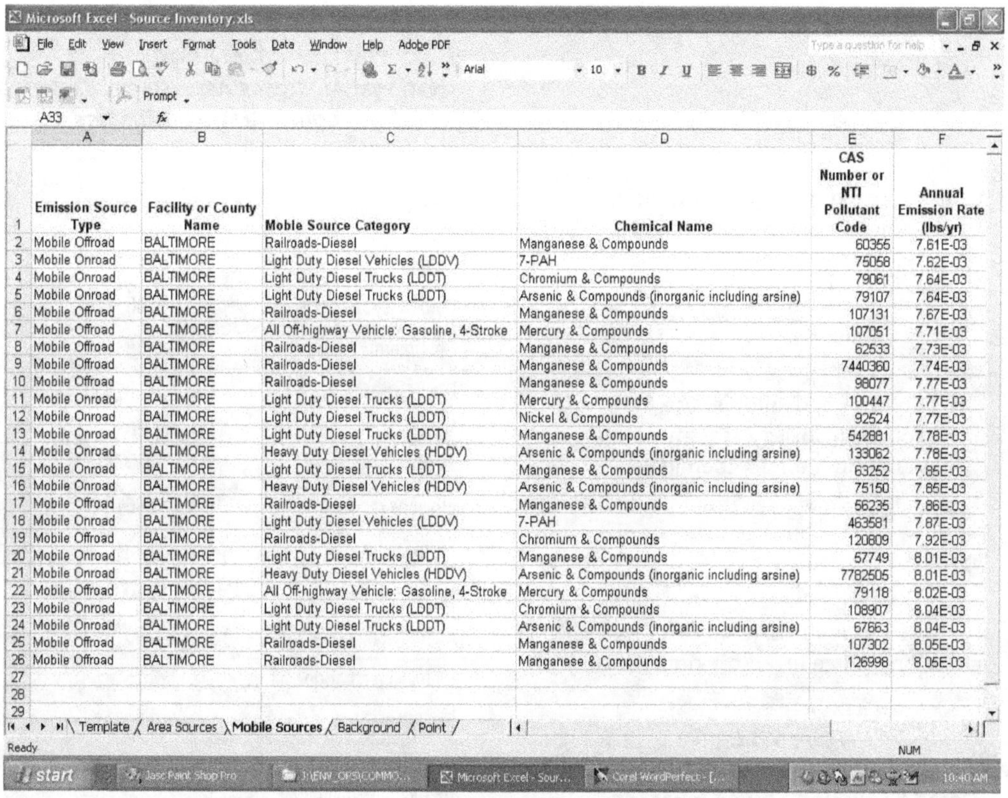

	A	B	C	D	E	F
1	Emission Source Type	Facility or County Name	Moble Source Category	Chemical Name	CAS Number or NTI Pollutant Code	Annual Emission Rate (lbs/yr)
2	Mobile Offroad	BALTIMORE	Railroads-Diesel	Manganese & Compounds	60355	7.61E-03
3	Mobile Onroad	BALTIMORE	Light Duty Diesel Vehicles (LDDV)	7-PAH	75058	7.62E-03
4	Mobile Onroad	BALTIMORE	Light Duty Diesel Trucks (LDDT)	Chromium & Compounds	79061	7.64E-03
5	Mobile Onroad	BALTIMORE	Light Duty Diesel Trucks (LDDT)	Arsenic & Compounds (inorganic including arsine)	79107	7.64E-03
6	Mobile Offroad	BALTIMORE	Railroads-Diesel	Manganese & Compounds	107131	7.67E-03
7	Mobile Offroad	BALTIMORE	All Off-highway Vehicle: Gasoline, 4-Stroke	Mercury & Compounds	107051	7.71E-03
8	Mobile Offroad	BALTIMORE	Railroads-Diesel	Manganese & Compounds	62533	7.73E-03
9	Mobile Offroad	BALTIMORE	Railroads-Diesel	Manganese & Compounds	7440360	7.74E-03
10	Mobile Offroad	BALTIMORE	Railroads-Diesel	Manganese & Compounds	98077	7.77E-03
11	Mobile Onroad	BALTIMORE	Light Duty Diesel Trucks (LDDT)	Mercury & Compounds	100447	7.77E-03
12	Mobile Onroad	BALTIMORE	Light Duty Diesel Trucks (LDDT)	Nickel & Compounds	92524	7.77E-03
13	Mobile Onroad	BALTIMORE	Light Duty Diesel Trucks (LDDT)	Manganese & Compounds	542881	7.78E-03
14	Mobile Onroad	BALTIMORE	Heavy Duty Diesel Vehicles (HDDV)	Arsenic & Compounds (inorganic including arsine)	133062	7.78E-03
15	Mobile Onroad	BALTIMORE	Light Duty Diesel Trucks (LDDT)	Manganese & Compounds	63252	7.85E-03
16	Mobile Onroad	BALTIMORE	Heavy Duty Diesel Vehicles (HDDV)	Arsenic & Compounds (inorganic including arsine)	75150	7.85E-03
17	Mobile Offroad	BALTIMORE	Railroads-Diesel	Manganese & Compounds	56235	7.86E-03
18	Mobile Onroad	BALTIMORE	Light Duty Diesel Vehicles (LDDV)	7-PAH	463581	7.87E-03
19	Mobile Offroad	BALTIMORE	Railroads-Diesel	Chromium & Compounds	120809	7.92E-03
20	Mobile Onroad	BALTIMORE	Light Duty Diesel Trucks (LDDT)	Manganese & Compounds	57749	8.01E-03
21	Mobile Onroad	BALTIMORE	Heavy Duty Diesel Vehicles (HDDV)	Arsenic & Compounds (inorganic including arsine)	7782505	8.01E-03
22	Mobile Offroad	BALTIMORE	All Off-highway Vehicle: Gasoline, 4-Stroke	Mercury & Compounds	79118	8.02E-03
23	Mobile Onroad	BALTIMORE	Light Duty Diesel Trucks (LDDT)	Chromium & Compounds	108907	8.04E-03
24	Mobile Onroad	BALTIMORE	Light Duty Diesel Trucks (LDDT)	Arsenic & Compounds (inorganic including arsine)	67663	8.04E-03
25	Mobile Offroad	BALTIMORE	Railroads-Diesel	Manganese & Compounds	107302	8.05E-03
26	Mobile Offroad	BALTIMORE	Railroads-Diesel	Manganese & Compounds	126998	8.05E-03
27						
28						
29						

Figure 9-10.
Sample Mobile Source Data in Emission Source Inventory Database

Step 8: Guidance for collecting information on background concentrations

Background concentrations are releases that are not the result of current human activity, including both natural and past human sources. Figure 9-11 shows the procedure for adding background concentration data to the Emission Source Inventory. The primary sources of background concentrations data are from state and local emission inventories and the National Air Toxics Assessment (NATA). Information about the NATA program is located at http://www.epa.gov.ttnatw01/nata/.

The NATA program provides background concentrations based on monitored values identified in the Cumulative Exposure Project (1990 study that estimated nationwide ambient concentrations of air toxics). Based on the study, the nationwide background concentration values developed for 13 toxic air pollutants are shown on Table 9-10 and are also available at http://www.epa.gov/ttnatw01/nata/haptbl.html.

The available data used by NATA are insufficient to address geographic variations in background, and the background concentrations are assumed to be constant across all census tracts. For pollutants whose background concentration values could not be identified in the technical literature, the background concentrations are assumed to be zero. Therefore, this may result in underestimation of outdoor concentrations for some toxic air pollutants. There is also some concern that, in certain circumstances, there is the potential for double counting. The Partnership should review the NATA data along with other emission data to determine whether the background concentration data are already accounted for by other emission data contained in the Emission Source Inventory.

Background concentration data tables are available for each state (on a county-wide basis) in Excel or PDF file format at http://www.epa.gov.ttn/atw/nata/tablcoc.html.

Scroll down the web page to the section labeled Download a State Summary Table and select your state. The background concentration data files are available in Excel spreadsheet and PDF file formats. The required information about stationary area sources to be included

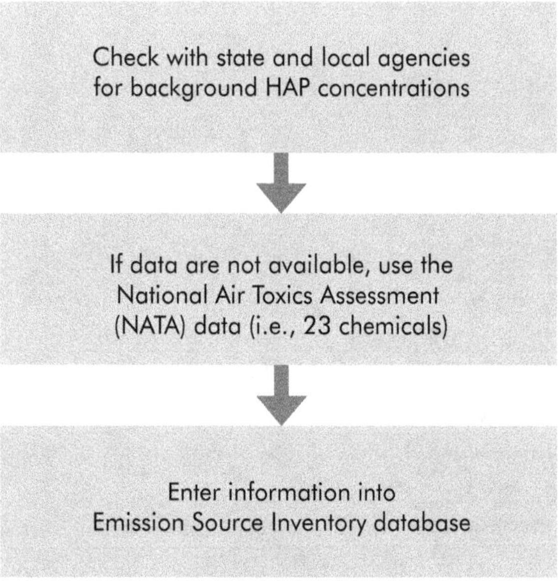

Check with state and local agencies for background HAP concentrations

↓

If data are not available, use the National Air Toxics Assessment (NATA) data (i.e., 23 chemicals)

↓

Enter information into Emission Source Inventory database

Figure 9-11.
Procedure for Adding Background Concentration Data to Emission Source Inventory

---

Table 9-10.
National Air Toxics Assessment Background Concentration Estimates

POLLUTANT	BACKGROUND CONCENTRATION ($\mu g/m^3$)
Benzene	0.48
Carbon tetrachloride	0.88
Chloroform	0.083
Dioxins/furans (toxicity equivalents)	0.000000015
Ethylene dibromide	0.0077
Ethylene dichloride	0.061
Formaldehyde	0.25
Hexachlorobenzene	0.000093
Mercury compounds	0.0015
Methylene chloride	0.15
Polychlorinated biphenyls	0.00038
Perchloroethylene (tetrachloroethylene)	0.14
Trichloroethylene	0.081

in the Emission Source Inventory, along with the corresponding data elements to be extracted from the NATA background concentration data tables, are provided in Table 9-11. Figure 9-12 shows an example of the Emission Source Inventory with sample background concentration data entered.

Step 9: *Guidance for collecting information on monitored concentrations*

This methodology encourages the use of ambient air monitoring data over concentrations estimated by modeling. However, when considering how to use monitoring data, it is very important to understand the goals of the monitoring program collecting the data and the limitations in the spatial and temporal coverage of the data. For example, a program designed to determine the average concentrations of air pollutants across a city might not be an appropriate surrogate for the concentration of that pollutant at the fenceline of a facility that emits it.

Several things should be considered when deciding whether or not to use monitoring data over estimated airborne pollutant concentrations.

- Does a monitoring network exist for the pollutants of interest and what are the averaging times of concern?
- Has the monitoring network been designed to locate points of maximum concentrations, average, other?
- Do the data set and analysis allow the impact of the most important individual sources to be identified if more than one source or emission point is involved?
- Is at least one full year of valid ambient data available ?

These questions are intended to help in making the decision whether to use monitoring data, modeling data, or both in the methodology. It is recommended that this discussion take place and the decision on the usefulness of available monitoring data be made prior to entering the data into the database (emissions inventory).

The procedure for adding monitored concentration data to the Emission Source Inventory is shown in Figure 9-13. The primary source the Partnership should use for information about monitored concentration data are state and local inventories. It is important to ensure that the inventory is set up so that monitoring data are clearly identified as such and cannot be mistaken for estimated concentration values derived from modeling. As discussed under step 3, state and local government authorities are responsible for permitting air releases and maintaining databases that contain information about local sources and their releases. These databases and associated files may contain monitored concentration data for the chemicals regulated by state and local legislation.

The Partnership can also obtain monitoring data from the AIRData website located at http://www.epa.gov/air/data/aqsdb.html.

The AIRData web site contains ambient concentrations of pollutants in outdoor air that are measured at more than 4,000 monitoring stations owned and operated mainly by state environmental agencies. They forward the hourly or daily measurements of pollutant concentration to EPA's database, and EPA computes a yearly summary for each monitoring station (maximum value, average value, number of measurements, etc.). AIRData has the yearly summary values only, and not the individual hourly or daily measurements.

Table 9-11.
Required Data Elements for Emission Source Inventory from NATA State Background Concentration Tables

EMISSION SOURCE INVENTORY DATABASE FIELD	FIELD NAME FROM NATA STATE BACKGROUND CONCENTRATION TABLE
Facility or County Name	County
Chemical Name	Pollutant
CAS Number	The Partnership will need to assign a CAS number to the chemical name because it is not provided in the NATA tables.
Estimated Ambient Air Concentration (μg/m3)	Estimated background

What is the next step after the Inventory Team has completed the first step of collecting information for the Emission Source Inventory database?

Information contained in the Emission Source Inventory can be summarized and displayed in a variety of ways that would be useful for the Partnership. Use of summary tables and figures helps to present large amounts of data in a form that can be readily understood. Examples include:

- Figure or map showing the study area with the major emission sources located
- Table showing the emission sources that were included and those not included in the study
- Table showing the chemicals that will be studied during the Initial, Secondary, and Final Screening steps

The goal for the Partnership, prior to starting the Initial Screen, will be to include as many chemicals and sources as possible in its inventory. The full Partnership committee should take part in reviewing the data in the Emission Source Inventory to ensure that the information is accurate and complete. The residents located in the study area can also assist in quality control of the source inventory by reviewing the database to make sure that all potential sources of emissions have been identified. Members of the Partnership can check the database through an on-the-ground check of the study area or by relying on the residents' knowledge of local facilities and their locations. Community residents and local businesses will be in the best position to judge the accuracy of the inventory database. Please see Chapter 8 in the Overview for suggestions on presenting the information in the inventory to the broader community.

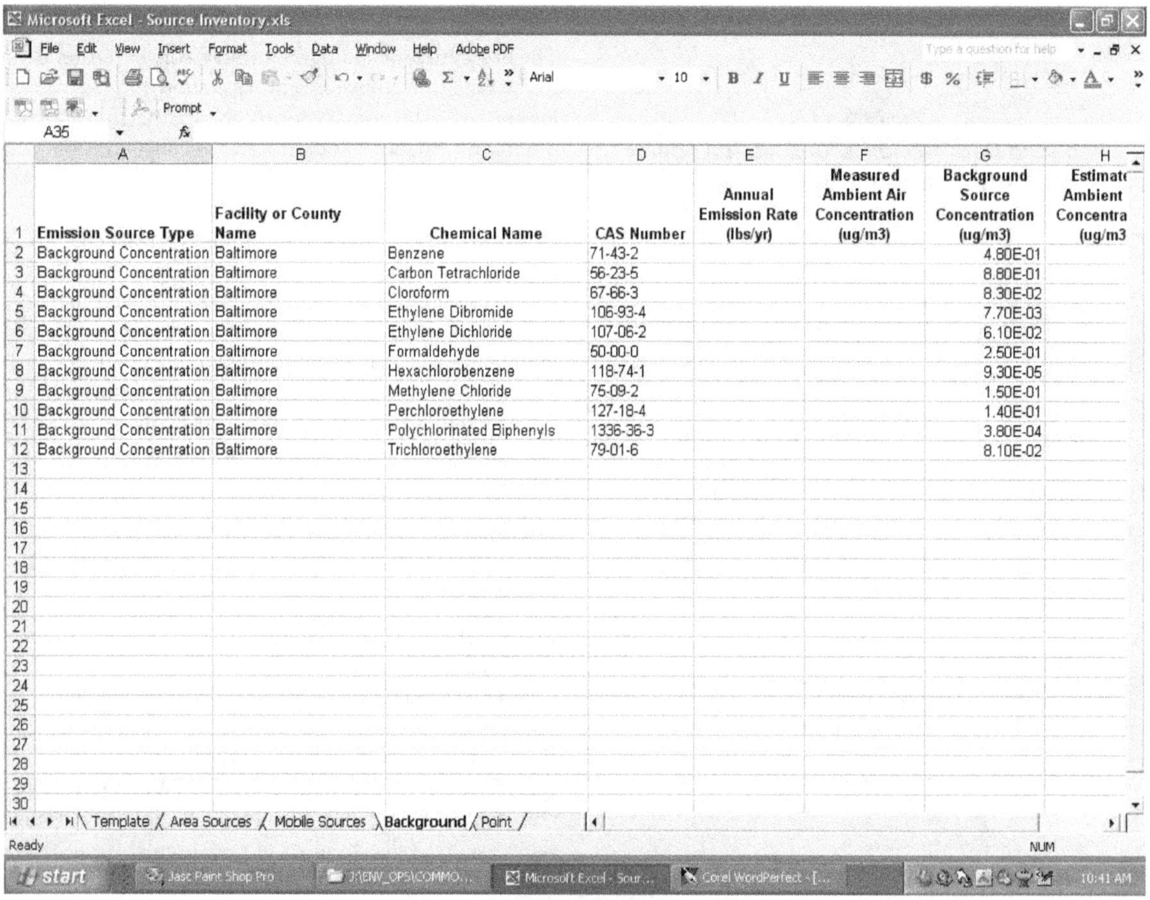

Emission Source Type	Facility or County Name	Chemical Name	CAS Number	Annual Emission Rate (lbs/yr)	Measured Ambient Air Concentration (ug/m3)	Background Source Concentration (ug/m3)	Estimat Ambient Concentra (ug/m3
Background Concentration	Baltimore	Benzene	71-43-2			4.80E-01	
Background Concentration	Baltimore	Carbon Tetrachloride	56-23-5			8.80E-01	
Background Concentration	Baltimore	Cloroform	67-66-3			8.30E-02	
Background Concentration	Baltimore	Ethylene Dibromide	106-93-4			7.70E-03	
Background Concentration	Baltimore	Ethylene Dichloride	107-06-2			6.10E-02	
Background Concentration	Baltimore	Formaldehyde	50-00-0			2.50E-01	
Background Concentration	Baltimore	Hexachlorobenzene	118-74-1			9.30E-05	
Background Concentration	Baltimore	Methylene Chloride	75-09-2			1.50E-01	
Background Concentration	Baltimore	Perchloroethylene	127-18-4			1.40E-01	
Background Concentration	Baltimore	Polychlorinated Biphenyls	1336-36-3			3.80E-04	
Background Concentration	Baltimore	Trichloroethylene	79-01-6			8.10E-02	

Figure 9-12.
Sample Background Concentration Data in Emission Source Inventory Database

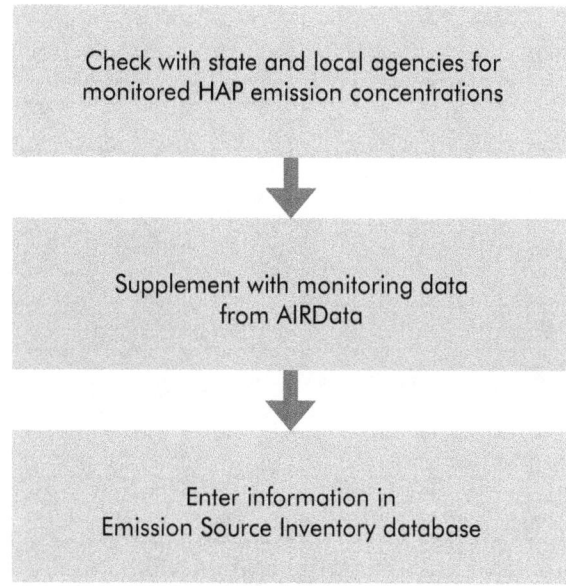

Figure 9-13.
Procedure for Adding Monitored Concentration
Data to Emission Source Inventory

Initial Screen

This section describes the steps the Initial Screen Technical Team can take to complete the Initial Screen. If resources are available, it is recommended that two technical teams be set up, one for estimating concentrations and one for developing screening concentrations. This will save time since these tasks can move forward simultaneously. Flow charts showing a summary of the steps for completing the Initial Screen are shown in Figure 10-1 and Figure 10-2. Figure 10-1 shows how to estimate ambient air concentrations for the chemicals in the Emission Source Inventory that will be reviewed in the Initial Screen. Figure 10-2 shows how to calculate the screening-level concentrations that will be used to identify community priorities.

Estimating Concentrations

Which of the sources will have their concentrations estimated in the Initial Screen?

During the Initial Screen ambient air concentrations for stationary point sources (with a receptor distance equal to or greater than 15 meters) and stationary area sources will be estimated. Releases from stationary point sources with a receptor distance less than 15 meters will not be reviewed in the Initial Screen but will be kept for review during the Secondary Screen. Mobile source emissions also will not be reviewed in the Initial Screen but will be kept for review during the Secondary Screen. Releases of mobile source chemicals from stationary point and area sources will also be reviewed in the Secondary Screen.

How will the technical team estimate these concentrations?

The technical team will use a look-up table, or a simple conversion factor, to convert emission rate information contained in the Emission Source Inventory to ambient air concentrations for each stationary point source and stationary area source. The estimated ambient air concentrations emitted from each stationary point source and stationary area source are added to the background concentrations contained in the Emission Source Inventory to obtain a total ambient air concentration for each chemical. The Partnership will then compare the total aggregate concentration for each chemical to the screening-level concentration to determine whether to keep the chemical for further analysis in the Secondary Screen.

How were the look-up table and the conversion factor developed and where can the technical team find a detailed description of the SCREEN3 air dispersion model?

The look-up table and the conversion factor were developed using the SCREEN3 air dispersion model. The look-up table was developed by running the SCREEN3 model using the input parameters shown in Table 10-1. The stack height and receptor distance values were varied prior to each of the modeling runs. The resulting look-up table is shown on Table 10-2. The Partnership can use the look-up table to estimate ambient air concentration by multiplying the emission rate (in lb/yr) by the unitized annual average concentration for the applicable stack height and receptor distance.

The conversion factor, 6.55 E-03 $\mu g/m^3$ per lb/yr, was developed by calculating the 95th percentile value from the look-up table. The estimated ambient air concentration is obtained by multiplying the emission rate (in lb/yr) by the conversion factor.

Step 1

Set up QA/QC procedures for data collection and entry and for calculations.

Step 2

Collect information on release type, stack height, and receptor distance for stationary point sources. (If the technical team has limited resources and chooses to use the conversion factor to estimate concentrations, skip this step.)

Step 3

Identify all the sources that are less than 15 meters from the nearest community exposure, and hold all the chemicals emitted from these sources for review in the Secondary Screen. They will not be reviewed in the Initial Screen.

Step 4

Identify all chemicals for mobile sources and hold these chemicals for review in the Secondary Screen. They will not be reviewed in the Initial Screen.

Step 5

For each chemical and for each of the stationary point sources releasing the chemical, estimate the concentration using the look-up table or conversion factor method. Add all the concentrations to get a total point source concentration.

Step 6

For each chemical released from stationary area sources, add the county-wide releases from all the area sources and estimate the concentration using the conversion factor.

Step 7

For each chemical, obtain the background concentration, if any.

Step 8

For each chemical, add the total stationary point source concentration, the total stationary area source concentration, and the background concentration, to obtain the total concentration from all sources.

Figure 10-1.
Procedure for Initial Screen

Step 1 - Set Community Risk Screening-Level Assumption Values

Cancer (C):
Risk screening level expressed as a statement of probability, e.g., 1E-06 (1 x 10^{-6} or 1 in 1 million).

Non-Cancer (N-C) Effects:
Risk screening level expressed as a ratio of exposure level to
hazard (toxicity) level or a hazard quotient (HQ), e.g., HQ = 1.

Step 2 - Decide on Toxicity Information Hierarchy

Technical team explains criteria for ranking sources of toxicity information that it will use to develop screening-level concentrations and presents a proposal for ranking sources to the Partnership for approval or modification.

Partnership **decides** on a method to rank the sources of information to guide the technical team.

Step 3 - Calculate an Air Exposure Screening-Level Concentration (SLC)

$$\text{Cancer (C) SLC } \mu g/m^3 = \frac{RSL}{UR}$$

where:

RSL = Cancer risk screening level (e.g., 1.0E-06)
UR = Chemical-specific inhalation unit risk estimate (per $\mu g/m^3$)

$$\text{Non-Cancer (N-C) SLC } \mu g/m^3 = SHQ * RfC * 100 \ \mu g/mg$$

where:

SHQ = Screening hazard quotient (e.g., 1.0)
RfC = Chemical-specific inhalation reference concentration ($\mu g/m^3$)

Figure 10-2.
Steps for Developing a Screening-Level Concentration

The conversion factor was selected by using a combination of stack height and distance to receptor that would give a concentration higher than 95% of all other combinations of stack height and receptor distance. This conversion factor is intended to be conservative enough that if it is used as a default value by a community that does not know stack heights and receptor distances, the concentration estimated will most likely be higher than would be estimated using measured stack heights and receptor distances. The conversion factor is intended to provide a level of protection in the Initial Screen when exact values for stack height and receptor distance are not known.

Detailed information on the SCREEN3 model can be found in the SCREEN3 user's guide. The user's guide can be downloaded from the EPA web site at http://www.epa.gov/scram001/userg/screen/screen3d.pdf.

Table 10-1.
SCREEN3 Input Parameters Used to Generate Initial Screen Look-Up Table

SCREEN3 MODEL PARAMETERS	VALUE	SCREEN3 MODEL PARAMETERS	VALUE
Emission source type	point	Receptor height	0
Emission rate	1 g/s	Landscape type	urban
Stack height	variable	Meteorological conditions	general full meteorology
Stack diameter	0.15 m	Building downwash	No
Stack gas exit velocity	2 m/s	Terrain calculation	No
Exit gas temperature	293 K	Receptor distance	variable
Ambient temperature	293 K		

Which method, the look-up table or the conversion factor, will the Partnership use to estimate concentrations?

The decision to use the look-up table or the conversion factor to estimate ambient air concentrations is based on the resources available to the technical team. The look-up table is the recommended method for estimating concentrations, but if the team's resources are limited, then the conversion factor can be used instead. To use the look-up table, the technical team will need to have information on stack height and distance to receptor for each source. If the resources the Partnership has allocated to this project are limited, and stack heights and receptor distance values are not easily obtained, the Partnership may elect to use the conversion factor instead to estimate ambient air concentrations.

The conversion factor will conservatively estimate ambient air concentrations approximately 95% of the time, as long as the study area contains a diverse array of stack heights and receptor distances. However, in certain situations (approximately 5% of the time), use of the conversion factor may underestimate the ambient air concentration for a chemical. This would occur, for example, if all of the sources in the study area have stack heights of 3 meters and all of the receptors are less than 50 meters from these sources. On the other hand, if all of the sources have a stack height of 30 meters and all of the receptors are greater than 50 meters from these sources, use of the conversion factor would overestimate the ambient air concentration for a chemical.

Use of the look-up table will typically result in lower concentration estimates than would be obtained using the conversion factor. Therefore, use of the look-up table is likely to screen out more chemicals during the Initial Screen and reduce the number of chemicals kept for further study in the Secondary Screen. Another advantage of this approach is that because of the work required for the Partnership to collect information about the stack height and distance to receptor needed to use the look-up table, there is a better chance that sources that are less than 15 meters to the closest receptor will be identified for closer review in the Secondary Screen.

Table 10-2.
Look-Up Table. Unitized Annual Average Concentration at Different Distances from a Stack
($\mu g/m^3$ per lb/year)

Stack Height (m)	Distance to Receptor							
	15 m	20 m	50 m	100 m	200 m	300 m	400 m	500 m
3	1.59E-02	1.55E-02	1.04E-02	3.97E-03	1.19E-03	5.78E-04	3.50E-04	2.39E-04
5	5.63E-03	7.05E-03	5.42E-03	3.33E-03	1.13E-03	5.64E-04	3.45E-04	2.37E-04
8	1.05E-03	2.20E-03	2.93E-03	2.23E-03	1.01E-03	5.33E-04	3.33E-04	2.31E-04
10	2.35E-04	9.51E-04	1.94E-03	1.57E-03	9.12E-04	5.08E-04	3.23E-04	2.26E-04
20	1.70E-09	1.08E-06	4.05E-04	4.59E-04	3.37E-04	2.79E-04	2.06E-04	1.55E-04
30	-	-	6.98E-05	2.10E-04	1.76E-04	1.30E-04	1.22E-04	1.04E-04
40	-	-	6.34E-06	1.02E-04	1.02E-04	8.53E-05	6.49E-05	6.49E-05
50	-	-	3.00E-07	4.79E-05	7.03E-05	6.08E-05	4.94E-05	3.81E-05
75	-	-	-	3.74E-06	2.95E-05	2.95E-05	2.54E-05	2.36E-05
100	-	-	-	1.11E-07	1.28E-05	1.63E-05	1.59E-05	1.41E-05

How will the Partnership collect the information needed to use the look-up table?

The Initial Screen Technical Team will work with and rely on the Partnership's Inventory Team and the community to collect information on stack height and distance to closest receptor for all stationary point sources. The work that the Inventory Team has done to collect release and location information for all stationary point sources to begin the Partnership's Emission Source Inventory database will provide a good foundation for collecting the information that will be needed for the Initial Screen. See step 2 below for more details.

What are the steps that the technical team will need to complete to estimate concentrations for the Initial Screen?

Step 1: Set up QA/QC procedures for data collection and entry and for calculations.

Step 2: Collect information on release type, stack height, and receptor distance for stationary point sources. (If the technical team has limited resources and chooses to use the conversion factor to estimate concentrations, skip this step.)

Step 3: Identify all the sources that are less than 15 meters from the nearest community exposure, and hold all the chemicals emitted from these sources for review in the Secondary Screen. They will not be reviewed in the Initial Screen.

Step 4: Identify all chemicals for mobile sources and hold these chemicals for review in the Secondary Screen. They will not be reviewed in the Initial Screen.

Step 5: For each chemical and for each of the stationary point sources releasing the chemical, estimate the concentration using the look-up table or conversion factor method. Add all the concentrations to get a total point source concentration.

Step 6: For each chemical released from stationary area sources, add the county-wide releases from all the area sources and estimate the

concentration using the conversion factor.

Step 7: For each chemical, obtain the background concentration, if any.

Step 8: For each chemical, add the total stationary point source concentration, the total stationary area source concentration, and the background concentration to obtain the total concentration from all sources.

How will the Partnership carry out these steps?

Step 1: Set up QA/QC procedures for data collection and entry and for calculations.

See discussion of QA/QC procedures in Chapter 9 of the Technical Guidance section of the Manual.

Step 2: Collect information on release type, stack height, and receptor distance for stationary point sources. (If the technical team has limited resources and chooses to use the conversion factor to estimate concentrations, skip this step.)

The technical team needs to collect and enter information about release type, stack height, and receptor distance into the Emission Source Inventory for each stationary point source. Three additional fields need to be added to the Emission Source Inventory database: (1) Release Type (stack or fugitive), (2) Stack Height (meters), and (3) Distance to Receptor (meters) to store this information.

Stationary point sources are designated as being either a stack or fugitive release type. The fugitive release type designation applies only to stationary point source emissions that are not released from stacks, vents, ducts, pipes, or other kind of confined stream. The Inventory Team can rely on the technical expertise of its members or on site visits to determine if the releases from a stationary point source are stack or fugitive.

Stack height information for stationary point sources can be obtained from the National Emission Inventory (NEI) database. Table 10-3 shows the appropriate NEI data table and field name to obtain the required stack height information for stationary point sources. Some sources handled as point sources in this Manual may be treated as area sources in the NEI database and will not be included in the NEI point source database. (Methods for turning the area sources into point sources and

Table 10-3.
Required Data for Emission Source Inventory from NEI Point Source Files

EMISSION SOURCE INVENTORY DATABASE FIELD	TABLE AND FIELD NAME FROM NEI POINT SOURCE DATA FILE
Stack Height	tblPointER –> sngStackHeight

Table 10-4.
Mobile Source Chemicals to Be Passed to the Secondary Screen

CHEMICAL NAME
1,3-Butadiene
Acetaldehyde
Acrolein
Arsenic compounds
Benzene
Beryllium compounds
Cadmium compounds
Chromium compounds
Dioxins/Furans as 2,3,7,8-TCDD TEQs
Ethylbenzene
Formaldehyde
Lead compounds
Manganese compounds
Mercury compounds
Methyl tertiary butyl ether (MTBE)
n-Hexane
Nickel compounds
Polycyclic organic matter (POM) as 7-Polycyclic aromatic hydrocarbon (PAH)
Polycyclic organic matter (POM) as 16-Polycyclic aromatic hydrocarbon (PAH)
Priopionaldehyde
Styrene
Toluene
Xylene

Source: Table 2. Mobile Source HAPs in the 1996 NTI, page 10, at http://www.epa.gov/ttn/chief/nti/ntipap1.pdf.

estimating an emission rate for each of these sources were provided in Chapter 9.) Since they are not included in the NEI point source database, stack height information will not be available from NEI. The technical teams will need to collect or estimate stack height information for each of these sources. Fugitive stationary point source emissions are assigned a default stack height value of three meters in the Emission Source Inventory.

Information about the distance to the nearest receptor can be obtained from a variety of sources, including small-scale maps, desktop mapping software, community members, and businesses, or, if necessary, the technical team can organize site visits to collect this information.

Step 3: *Identify all the sources that are less than 15 meters from the nearest community exposure, and remove all the chemicals emitted from these sources from the review. These chemicals will be reviewed in the Secondary Screen.*

Stationary point sources, and associated chemicals, with receptor distances that are closer than 15 meters will be passed to the Secondary Screen for review. If the look-up table is used to estimate concentrations, the information about the distance to receptor for each source was collected and added to the Emission Source Inventory during step 2. If the conversion factor is used to estimate concentrations, the Inventory Team will need to determine which sources are closer than 15 meters to the nearest community exposure.

Step 4: *Remove all the chemicals from mobile sources. These chemicals will also be reviewed in the Secondary Screen.*

Mobile source emissions will not be reviewed in the Initial Screen but will be kept for review during the

Secondary Screen. A list of the mobile source chemicals is shown in Table 10-4. If any of the mobile source chemicals shown in Table 10-4 are released from either stationary point sources or stationary area sources then these releases will also be passed to the Secondary Screen for review.

Step 5: *For each chemical and for each of the stationary point sources releasing the chemical, estimate the concentration using the look-up table or conversion factor method. Add all the concentrations to get a total point source concentration.*

For stationary point sources when the stack height and the distance to the receptor is known, the look-up table is used to estimate the ambient air concentration. This is accomplished by selecting the appropriate row on the table for the stack height and the appropriate column for the receptor distance and obtaining the corresponding unitized annual average concentration (in units of $\mu g/m^3$ per lb/yr). Then, the estimated ambient air concentration is obtained by multiplying the emission rate (in units of lbs/yr) by the unitized annual average concentration for the applicable stack height and receptor distance. For fugitive releases, a default value of 3 meters is used for the stack height. This process is repeated to estimate the concentration of each record in the Emission Source Inventory that contains stack height, receptor distance, and emission rate information.

When the stack height and distance to receptor are not known for a stationary point source, or if there are insufficient Partnership resources available to collect this information, the conversion factor of 6.55 E-03 $\mu g/m^3$ per lb/yr can be used to estimate the ambient air concentrations. The estimated ambient air concentration is obtained by multiplying the emission rate (in units of lbs/yr) by the conversion factor.

The estimated ambient air concentrations are entered into the Estimated Ambient Air Concentration field in the Emission Source Inventory. For stationary point sources, a separate value for the estimated ambient air concentration (obtained using the look-up table or the conversion factor) will be entered into the Estimated Ambient Air Concentration field in the Emission Source Inventory for each chemical. Add all the concentrations to get a total point source concentration for each chemical.

Step 6: *For each chemical released from stationary area sources, add the county-wide releases from all the area sources and estimate the concentration using the conversion factor.*

To estimate the ambient air concentrations for stationary area sources, the conversion factor of 6.55 E-03 $\mu g/m^3$ per lb/yr is used. The emission rates for a given chemical are summed for all of the stationary area sources. Then the estimated ambient air concentration is obtained by multiplying the sum of all the emission rates (in units of lbs/yr) for each chemical by the conversion factor. The estimated ambient air concentrations are entered into the Estimated Ambient Air Concentration field in the emission source inventory for each chemical.

Step 7: *For each chemical, obtain the background concentration, if any.*

Background concentrations for chemicals in the study area have been collected and entered into the Emission Source Inventory by the Partnership prior to the Initial Screening step. However, additional relevant monitoring data may be collected and entered.

Step 8: *For each chemical, add the total stationary point source concentration, the total stationary area source concentration, and the background concentration, to obtain the total concentration from all sources.*

Add the total stationary point source concentration (step 4), the total stationary area source concentration (step 5), and the background concentration (step 6) to obtain the total concentration from all sources. The total ambient air concentration is then compared to the screening-level concentration for each chemical. Guidance for developing screening-level concentrations is provided below.

Developing Screening-Level Concentrations

How will the Screening-Level Concentration Technical Team work with the Partnership to set the screening-level concentrations?

To ensure that estimates of screening-level concentrations are consistent with the goals and values of the Partnership, the technical team will need to work closely with the overall Partnership to complete its work. Decisions on the risk screening level and sources of toxicity information need to be made, as described in the Overview, by the full Partnership committee. The technical team will help to provide the Partnership with the information and background that it will need to make these decisions. Once the full Partnership committee has made the necessary decisions, the technical team will calculate an air exposure screening-level concentration for each community chemical.

What steps will the technical team need to complete to develop screening-level concentrations for the Partnership?

The technical team will, in general, assist the Partnership in providing the education that its members will need to participate fully in decisions that the Partnership will need to make to ensure that the work to set screening-level concentrations meets its goals and values. In particular, the Screening-Level Concentration Technical Team will need to complete the following steps:

Step 1: Provide assistance and background education to help the Partnership decide on the risk screening levels that it will use to identify the community's priority chemicals that will be evaluated for possible community action.

Step 2: Explain criteria for ranking sources of toxicity information that the Partnership will use to develop its screening-level concentrations. Present a proposal for ranking sources to the Partnership for approval or modification to help the Partnership decide on a method to

rank the sources of information to guide the technical team. See the Overview for a discussion of the choice of toxicity sources. Appendix H provides additional information on types and sources of toxicity values needed to complete this step.

Step 3: Once the Partnership has decided on risk screening levels and on a plan for accessing toxicity information, develop air exposure screening-level concentrations for the Partnership and store them in the Partnership database for the Partnership to use. Appendix I provides detailed instructions for calculating screening-level concentrations.

Figure 10-2 summarizes these steps. Steps 1 and 2 of Figure 10-2 list the risk screening level concentration assumptions and toxicity information hierarchy that are required. Step 3 of Figure 10-2 provides the mathematical relationships the technical team can use to calculate screening-level concentrations, one for each of the two types of long-term toxicity potentially associated with each chemical. For detailed information and guidance for each of these steps, see the appendices listed above.

What are the next steps after the technical team has developed screening-level concentrations for the Partnership?

The technical team will enter a screening-level concentration for each community chemical into the Partnership's inventory database. It will provide the Partnership with a complete description of the method and information used to develop screening-level concentrations so that the Partnership and community can understand all of the assumptions and information that were included in calculations for each of its screening-level concentration values. The Partnership may want to develop an indication of the level of uncertainty for each screening-level concentration value to summarize the quality of the information used to develop the screening-level concentrations. Then, working together, the technical team that estimated the chemical concentrations in community air and the Screening-Level Concentration Technical Team will develop a report for the Partnership that summarizes results from comparing community chemical screening-level concentrations to the estimated ambient air concentrations. The Partnership will use results of these comparisons to identify chemicals that may need further analysis in the next step of the screening process.

Secondary Screen

What is the overall goal for the technical team?

The goal of the Secondary Screening process is to (1) provide a more refined estimate of the ambient chemical concentrations for chemicals remaining after the Initial Screening process, and (2) assess chemicals released from mobile sources. The results of the Secondary Screening process will be compared to the same screening-level concentrations developed during the Initial Screen to identify those chemicals that require further review in the Final Screen. A flow chart summarizing the steps for completing the Secondary Screen is shown in Figure 11-1.

How will the technical team estimate concentrations?

The technical team will estimate chemical concentrations by using the Industrial Source Complex Short Term (ISCST) model. This model conforms with EPA's *Guideline on Air Quality Models (Revised)* and allows the technical team to model emissions from a wide range of sources, employing the straight-line, steady-state Gaussian plume equation to estimate ambient concentration values. The ability of the mathematical approach used by ISCST, given accurate input data, to accurately estimate airborne concentrations of a chemical has been investigated in several studies. In one extensive investigation (EPA 1982. *Evaluation of Rural Air Quality Simulation Models.* EPA-450/4-83-003), estimated concentrations of SO_2 around a power plant derived using the same algorithms that ISCST uses were compared to actual measured concentrations over a two-year period. Comparison of the 25 highest estimated values and the 25 highest measured values indicated overprediction. Estimated values were generally within a factor of 2 of those measured. Prediction accuracy from one location to another within the receptor grid was variable. Results for different meteorological conditions indicated that the model tended to shift to underprediction as wind speed increased. Accuracy was also influenced by atmospheric stability conditions. For class A and B (unstable atmospheric conditions) the model overpredicted, while for class E and F (stable) the model underpredicted. There was a general trend toward model underprediction as the averaging period increased (i.e., 1-hour, 3-hour, 24-hour).

Why not use NATA's national-scale assessment of 33 air pollutants?

As a part of EPA's National Air Toxics Assessment (NATA), EPA has conducted a national-scale assessment of 33 air pollutants. This assessment used the ASPEN air dispersion model to estimate air concentrations for these 33 pollutants for each census tract in the nation. These concentrations are available on the web at http://www.eoa.gov/ttnatw01/nata.

Stationary area and mobile source concentrations were estimated for the national-scale assessment using the pseudo-point method similar to the one recommended in this Manual. Since the methods used in the national-scale assessment and this Manual are similar, the possibility of saving Partnership resources by using the readily available national-scale assessment concentrations needs to be considered. This Manual recommends that the Partnership do its own estimation of stationary area sources and mobile sources using the pseudo-point method for three reasons. First, while access to the census tract concentrations for 33 air pollutants is available, the concentrations for each source category type may not be easily available. For example, the national-scale assessment concentrations for area sources may include some sources that are categorized as point sources in this Manual. As a result, using the area source concentrations from the national-scale assessment would result in double counting of some sources. Second, the national-scale assessment is conducted once every three years, so the Partnership may have access through its state partner to more current release information for its modeling. And third, the national scale assessment estimated concentrations for only 33 chemicals and the Partnership's inventory may have mobile or stationary area source chemicals that are not included in this list of 33 chemicals.

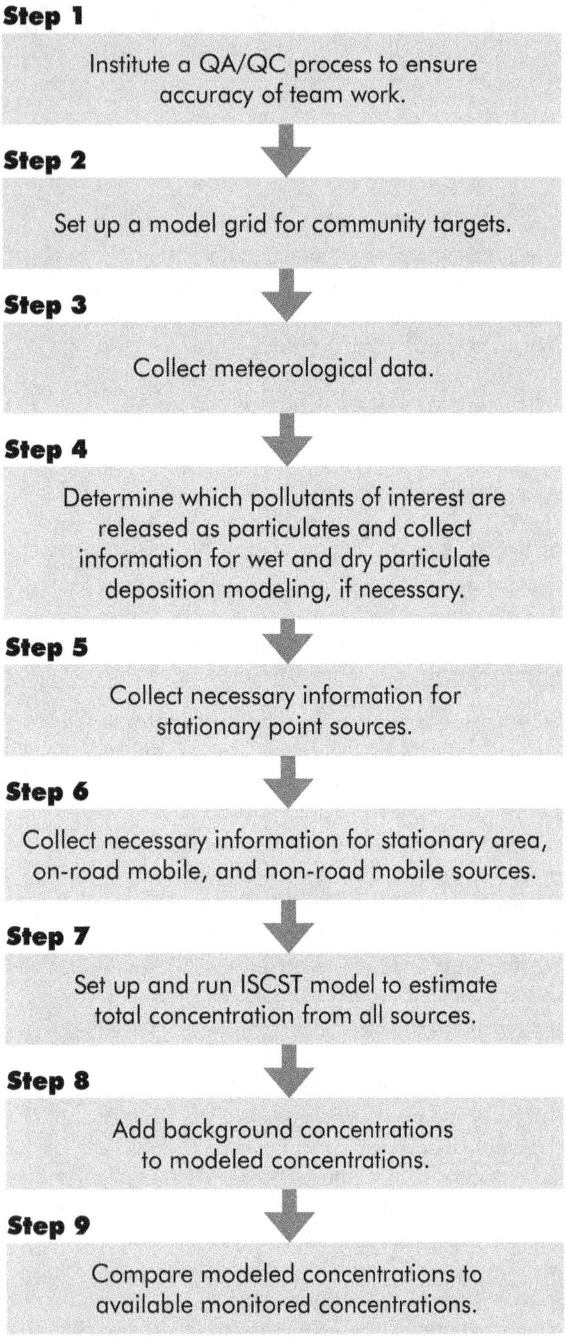

Step 1

Institute a QA/QC process to ensure accuracy of team work.

Step 2

Set up a model grid for community targets.

Step 3

Collect meteorological data.

Step 4

Determine which pollutants of interest are released as particulates and collect information for wet and dry particulate deposition modeling, if necessary.

Step 5

Collect necessary information for stationary point sources.

Step 6

Collect necessary information for stationary area, on-road mobile, and non-road mobile sources.

Step 7

Set up and run ISCST model to estimate total concentration from all sources.

Step 8

Add background concentrations to modeled concentrations.

Step 9

Compare modeled concentrations to available monitored concentrations.

Figure 11-1.
Procedure for Secondary Screen

Where can the technical team find a detailed description of the ISCST model?

The most recent version of ISCST can be obtained for free from EPA's Technology Transfer Network (TTN), Support Center for Regulatory Air Models (SCRAM) web site at http://www.epa.gov/ttn/scram, under the Dispersion Models link. The hardware and operating system requirements can be found in volume 1 of the user's guide. However, most modern computers using the Windows operating system can run ISCST.

There are several commercial versions of ISCST that can be obtained from vendors. These versions of ISCST are designed to run on Windows operating systems and may be easier to use and provide more features for the technical team. These Windows-based programs can simplify the model setup process, construction of necessary input files, and interpretation of results. While the actual savings of resources will differ from case to case, it will likely increase as the number of emissions and receptors increase and other advanced options are selected. However, it should be noted that these versions of ISCSTcan cost a substantial amount (approximately $1,250-1,500), and it is left to the Partnership to determine whether it would be worthwhile to invest in such software.

The ISCST3 model will also be available for use via the Internet using the Internet Geographic Exposure Modeling System (IGEMS), a program under development by EPA. But a fairly new computer with high-speed Internet access is needed to run the model. IGEMS also includes some of the data needed to run the model and will include GIS and the ability to perform risk assessment calculations in the near future. Information on IGEMS can be accessed at http://www.epa.gov/oppt/exposure/

What additional information will the technical team need to gather for the Emission Source Inventory?

Some of the information about the sources in the study needed to run the ISCST model (i.e., emission rate and stack height) were collected and entered into the Emission Source Inventory during the Initial Screen. In addition, the following new information is needed to run the ISCST model:

- Location of Potential Receptors
- Meteorological Data
- Information for Wet and Dry Particulate Deposition
- Stationary Point Source Parameters
- Information for Modeling Stationary Area and Mobile Source Emissions

The sections below discuss the various types of information required to run ISCST and where the technical team can find this information. The technical team should consult volume 1 of the ISCST user's guide on how to format this information so the model can use it properly. If commercial software is being used, the technical team should consult the software user support documents. It is recommended that the technical team work with the Inventory Team to collect the required information. The Inventory Team should provide support to the technical team throughout the process, providing emissions data and other source-related data as necessary. This will permit the technical team to focus on setting up and running the model and avoid potential duplication of information-gathering efforts.

Location of Potential Receptors

The ISCST data input file requires the technical team to specify a receptor network. This network can be input using a Cartesian (grid) network or using a polar coordinate network, although the Cartesian system appears to be the method of choice, primarily because it is compatible with the latitude and longitude or Universal Transverse Mercator (UTM) coordinates given for the locations of sources and receptors.

The grid network should be developed to identify both the sources of chemical emissions and the extent of the study area. This point of reference will subsequently be used to describe the locations of all sources and receptors. The technical team should determine a point of reference on the grid, also known as the location of origin in the study area. The technical team should also identify the locations of sensitive populations (i.e., schools, hospitals, nursing homes, residential communities, etc.) so that concentrations can be estimated for these sites. The key values that the technical team should determine for this grid are

- Total length of grid
- Total width of grid
- Spacing along length
- Spacing along width

- Elevation at each grid location (if elevated terrain or flagpole receptors are considered)

Distance and height entries should be provided in units of meters. It should be noted that this information is specific for Cartesian grid networks and that the technical team should consult volume 1 of the ISCST user's guide (or commercial software support documents) for more detailed information on setting up polar coordinate systems if that system is used.

Meteorological Data

ISCST requires at least one year of hourly meteorological data contained in a data file formatted to work with the model. But the technical team should collect five years of hourly meteorological data to account for year-to-year variations that have been formatted to work with the ISCST model. Meteorological data are available for weather stations located throughout the country, so the technical team should select a weather station that is representative of the study area. The technical team must also determine the anemometer height of the weather station being used for the meteorological data and mixing height information. Lastly, the technical team will have to classify the study area setting as either urban or rural.

Information for Wet and Dry Particulate Deposition Modeling

Additional data are required to model the deposition of particulates. This information includes additional meteorological data as well as information on the physical characteristics of the particulates being modeled.

Stationary Point Source Parameters

The technical team will be able to use some of the data from the Emission Source Inventory collected during the Initial Screen. However, additional data about the sources will be required. This additional data will vary, depending on the type of source being modeled, and are discussed as follows.

Stationary Point Stack

Information on emission rate (in grams per second) and stack height (in meters) for each source will have been collected during the Initial Screen to model stationary point sources. The Technical Team will need to collect the following additional information about the sources for the Secondary Screen:

- Stack diameter, in meters
- Stack exit velocity, in meters per second
- Stack exit temperature, in degrees Kelvin
- Source location, relative to the origin location, in meters

Stationary Point Fugitive

Since fugitive emissions are uncontrolled and not necessarily associated with a stack, the exact location of the fugitive emissions may or may not be known. For simplicity's sake, the fugitive emissions of a chemical at a given facility can be combined and modeled. This approach is relatively conservative because combining all fugitive releases into a single point source results in higher source strength and concentrations than if they were modeled as an area source, where the same mass of pollutant was released over a larger area, resulting in a lower source strength.

For each facility stationary fugitive sources will be modeled as a stationary point source, with the ISCST source-specific inputs set to represent no plume rise, using the following values:

- Stack height: 5 meters
- Stack diameter: 0.001 meters
- Stack exit velocity: 0.001 meters/second
- Stack exit temperature: 298 Kelvin

These parameters are chosen to represent releases relatively close to the ground that do not rise because of their elevated temperature or high exit velocity. This conservative approach is expected to be more protective than if greater release heights, temperatures, and exit velocities were chosen.

Information for Modeling Stationary Area and Mobile Source Emissions

Stationary area and mobile sources are modeled following a simplified version of the methodology from EPA Region 6 Regional Air Impact Modeling Initiative (RAIMI 2001). This method distributes stationary area and mobile source emissions evenly over the area of the census tracts in the study area. To use this approach the technical team will need to apportion county-wide mobile on-road and non-road mobile and stationary area source emission rates, collected from National, state, and local databases, to the census tracts in the study area.

The technical team will then apportion the census tract emission rates to five pseudo-stack points located in or near each census tract. The pseudo-stacks will be run in ISCST as point source emissions. One point will be located at the center of the census tract, while the remaining four will be located on the periphery, at the four different cardinal directions (i.e., North, South, East, and West). Detailed information for apportioning county wide emissions and locating pseudo-stack points are provided in Step 6 below. The following information will need to be collected for each stationary area and mobile source (the county wide emission rates that will be used were collected earlier in screening process):

- Information needed to apportion county wide emission rates to census tracts
- Emission rate in g/s apportioned to each census tract
- Location of centroids for census tracts in study area, relative to origin location, in meters
- Location of remaining four emission points, relative to origin location, in meters

As with stationary fugitive emissions, stack parameters will be set to represent no plume rise.

Information on the location of each census tract centroid in the partnership area and the area of each census tract can be obtained from the files contained in a tool called LandView. LandView reflects the collaborative efforts of the U.S. Environmental Protection Agency (EPA), the U.S. Census Bureau, the U.S. Geological Survey (USGS), and the National Oceanic and Atmospheric Administration (NOAA) to provide the public ready access to published Federal spatial and related data and is available for purchase. More information on its capabilities, availability and cost is available at: http://landview.census.gov/.

A resource available free to the public is the Census Bureau's American Fact Finder: http://factfinder.census.gov/.

This website provides access to much of the census and employment related information required for use apportioning county wide emission rates to census tracts. The IGEMS system being developed by EPA also includes census data: http://www.epa.gov/oppt/exposure/.

Are there special situations the technical team will need to consider in designing its modeling approach?

The technical team will need to initially determine whether the simplified ISCST modeling approach will be adequate for estimating the concentrations of pollutants in their community. To save resources, the simplified ISCST modeling approach should be used whenever possible, but in some cases more sophisticated modeling will be required. For example, if there are tall buildings near sources of emissions, then the technical team may want to go beyond the simple approach and consider building downwash in their modeling. The wind blowing around a building makes areas of rapid movement that are greater than if the building was not there. If release of a pollutant takes place from a roof vent or short stack, those releases can get caught in the areas of rapid movement. In the zone nearest the building, called the cavity zone, plumes can travel without much dilution and can be brought to the ground intermittently. This phenomenon can result in higher concentrations of the pollutant on the side of the building than would be predicted if the building was not present.

EPA provides guidance on whether building downwash should be considered when conducting air dispersion modeling. The decision is based largely on the height of the stack of concern in relation to the height or width of nearby buildings. The area of influence of a building is the lesser value of its height or width. Emissions within this area may be affected by downwash.

Stacks greater than EPA Good Engineering Practice stack height are generally not impacted by building effects, and downwash does not need to be considered for them. In its example for modeling urban air toxics, EPA used a 65-meter stack height as an upper limit for stacks potentially impacted by building effects. By referring to EPA guidance at the following web site, the Partnership can determine the appropriate stack height cutoff to use: http://www.epa.gov/scram001/guidance/guide/uatexample.pdf.

The technical team could use a simple approach to help determine when to consider including downwash in their modeling using local knowledge. The approach would be to consider the possibility of including building downwash for any stack less than 65 meters high with a building within 30 meters. This is based on an old "rule of thumb" that a stack next to a building must be at least 2.5 times the height of the building in order to avoid downwash of the plume, and on the area-of-influence guidance discussed above.

In addition to building downwash, unusual terrain can also change wind patterns that affect dispersion. If the study area is located in an area with a complex terrain, then more sophisticated modeling would be needed, for example, if the study area were located in a valley surrounded by high hills on hillsides higher than the sources of emission. The ISCST3 modeling approach used in support of the methodology is run without terrain influences, i.e., flat terrain is assumed. However, the ISCST3 model can adjust the plume heights by the receptor elevation above or below the stack base to account for the effects of elevated and complex terrain. The ISCST user's guide contains information for handling terrain.

Where the urban area is in mountainous terrain, terrain effects are important for sources with stacks. First, the impact of individual plumes on elevated terrain results in higher air concentration (through placing the receptor at the correct higher air concentration and vertical location within the plume and estimating the impact of the plume upon intervening terrain). Second, wind channeling due to terrain can cause higher air concentrations. The ISCST modeling approach does not address wind channeling effects other than if these effects are captured by the available meteorological data. If the area contains complex terrain features that are expected to significantly affect the modeled concentrations, a dispersion model that handles such situations should be selected from those listed in the EPA *Guideline on Air Quality Models*.

What are the steps the technical team will need to take to complete the work of estimating concentrations for the Secondary Screen?

After the technical team has decided whether or not to consider building downwash and terrain they should follow these steps to complete the Secondary Screen:

Step 1: Institute a QA/QC process to ensure accuracy of team work.

Step 2: Set up a model grid for community targets.

Step 3: Collect meteorological data.

Step 4: Determine which pollutants of interest are released as particulates and collect information for wet and dry particulate deposition modeling, if necessary.

Step 5: Collect necessary information for stationary point sources.

Step 6: Collect necessary information for stationary area, on-road mobile, and non-road mobile sources.

Step 7: Run the ISCST model to estimate total concentrations from all sources using five years of collected meteorological data.

Step 8: Add background concentrations to model concentrations.

Step 9: Compare modeled concentrations at the monitoring site to available monitored concentrations.

How will the technical team complete these steps?

Step 1: Institute a QA/QC process to ensure accuracy of team work.

Refer to the QA/QC section (step 1) in the Technical Guidance section.

Step 2: Set up a model grid for community targets.

The technical team needs to review the study area and determine how best to divide up the area to run in ISCST. Land use/land cover (LULC) maps, topographic maps, aerial photography, and GIS software are good sources of information. Elevation data can be obtained using 90-meter spaced USGS digital elevation data. The technical team should also identify the locations of sensitive populations (i.e., schools, hospitals, retirement communities, etc.) so that specific concentrations can be modeled for these sites. After surveying the study area, the technical team should lay out a grid network on a map of the study area. This will assist the technical team in determining the length and width of the grid network and the grid spacing along both axes, and will allow the technical team to choose a point of origin. Use of UTM coordinates can make it easier for the technical team to develop the grid network, as many databases use this to identify the location of a facility. Multiple receptor grids can be used to delineate the study area. Discrete receptor locations can also be entered into ISCST.

The technical team should keep in mind that the finer the grid spacing, the greater the time needed to run the ISCST model. The model calculates the sum of the user-specified hourly or annual average concentrations of a chemical from all sources defined on the grid, at each grid node or receptor location specified by the user. As the number of user-defined receptor sites and/or grid nodes increase, so do the number of calculations required to complete the modeling run.

A number of recent efforts to model urban air toxics at the community level have involved the use of grid spacing ranging from 100 meters to 1,000 meters. The use of a finer grid will allow the model to calculate concentrations at more grid nodes, which are spaced closer together, and potentially closer to a receptor of concern.

Step 3: Collect meteorological data

The technical team should collect five years of hourly meteorological data to account for year-to-year variations that have been formatted to work with the ISCST model. These data should include information on precipitation for use in modeling wet deposition of particulates. Meteorological data can be obtained from the National Oceanographic and Atmospheric Administration (NOAA) National Climatic Data Center (NCDC) web site at http://lwf.ncdc.noaa.gov/oa/ncdc.html; EPA's SCRAM web site at http://www.epa.gov/ttn/scram, under the Meteorological Data link, the Solar and Meteorological Surface Observation Network (SAMSON) compact disks, etc.

Meteorological data are also available for weather stations located throughout the country. If weather station data are used, the technical team should select a weather station that is either located inside the study area or representative of the study area. Data derived from these sources must then be run through a preprocessing program to put the data in the proper data file format for use with ISCST. Preprocessor programs are available at EPA's SCRAM web site under the Meteorological Data link.

The technical team must also determine the anemometer height of the weather station being used for the meteorological data. This information is available at the NCDC web site, under the Weather Station/City link. Additionally, information on the upper air meteorological conditions (i.e., mixing height data) for some states can be obtained from the EPA SCRAM web site at http://www.epa.gov/scram001/tt24.htm#mixing.

Mixing height data are also available for a fee from the National Climatic Data Center web site at http://www.ncdc.noaa/oancdc.html.

Lastly, the technical team will have to classify the study area setting as either urban or rural. Section 8.2.8 of EPA's *Guideline on Air Quality Models (Revised)* provides two procedures for determining if a study area is urban or rural. The more popular method (i.e., Land Use Procedure) looks at a 3 km circle around the study area. If 50% or more of the area in this 3 km circle can be classified as urban, then the study area should be classified as urban.

Step 4: *Determine which pollutants of interest are released as particulates and collect information for wet and dry particulate deposition modeling, if necessary.*

Particle deposition is a process in which particulates emitted from a source come into contact with a surface by the combined processes of atmospheric turbulence and gravitational settling. Particle deposition is evaluated during the Secondary Screen for exposure due to inhalation only. Emissions to air in the form of particles subject to deposition behave differently than gases. Deposition of particles may result in lower concentration in air at a distant receptor because the particles settle to the ground as they are carried downwind. This results in a reduction in exposure at a distant receptor, but may result in higher particulate concentrations for receptors near the source. Fugitive dust, mobile sources, metal smelters, coal-fired utilities, and wood stoves are examples of sources with particulate emissions.

In step 3, meteorological data were collected for ISCST modeling. In this step additional data required to model the deposition of particulates are collected. The technical team needs to determine if any of the pollutants of interest are released to air as particulates. If so, then additional information on the physical characteristics of the particulates being modeled, such as their size distribution and settling rates, should be collected. It is recommended that emissions of metals, including arsenic, beryllium, cadmium, chromium, lead, manganese, and nickel, be modeled as particulates.

For specific guidance on how to model deposition refer to the *ISC User's Guide volumes 1 and 2,* available through the Technology Transfer Network Support Center for Regulatory Air Models (TTN SCRAM), at http://www.epa.gov/scram001/userg/regmod/isc3v1.pdf and http://www.epa.gov/scram001/userg/regmod/isc3v2.pdf.

For general guidance see the Guidance/Support page at http://www.epa.gov/scram001/tt25.htm#guidance.

For an example of how deposition is modeled see *Example Application of Modeling Toxic Air Pollutants in Urban Areas* available at http://www.epa.gov/scram001/guidance/guide/uatexample.pdf.

Step 5: *Collect necessary information for stationary point sources (stack and fugitive).*

The technical team will start by using data from the Emission Source Inventory that were developed for use in the Initial Screen. It is recommended that the technical team work with the Inventory Team to collect the new information that will be needed. The technical team should then contact state and local agencies to obtain more detailed information on stack parameters for stationary point sources. National databases (such as NEI and TRI) can be used to supplement missing information. If the required information is still not available, the technical team may need to contact facility representatives or use conservative default information. It may also be necessary to visit the site and make visual estimates.

Fugitive emissions for each chemical at a facility are modeled from a single stack with default characteristics described elsewhere. Each pseudo-stack needs to be assigned a location on the modeling grid. The technical team may want to consider the following to choose a location for the pseudo-stack:

- Location of operations at the facility with fugitive emissions
- Location of sensitive populations
- Direction of prevailing winds
- Distance from operations with fugitive emissions to fenceline
- Distance from facility fenceline to populated area

Using the grid network developed in step 2, the technical team should identify the stationary point source (i.e., stack location) on the map, and estimate the location of the point source relative to the origin of the grid network. Again, using UTM coordinates can make this step easier for the technical team.

Keep in mind that for stationary sources these new estimates of concentrations developed by modeling are still designed to be conservative, that is, overestimates. To ensure that the Secondary Screen is conservative, the Partnership and its technical team will still be using the readily available data on maximum permitted release amounts as input to the model. If the release data available to a community Partnership only have estimated actual release amounts and not maximum permitted release amounts for stationary sources, the Partnership will need to increase these release estimates, possibly multiplying the release amount by a factor agreed on by the Partnership, to ensure that the Secondary Screen is conservative. Release estimates taken from the TRI database, for example, are estimated actual releases, not maximum permitted releases, so the Partnership's technical team will need to increase these amounts to ensure that the Secondary Screen is conservative.

Step 6: *Collect necessary information for stationary area, on-road mobile, and non-road mobile sources.*

County-wide emission rate data for on-road mobile, non-road mobile, and stationary area sources will be apportioned to census tract emission rates. To estimate emissions of these types of sources, spatial **allocation** or **apportioning** is used to estimate emissions at a small geographic scale based on national emissions estimates. County-wide emission rate data for on-road and non-road mobile sources can be obtained from state and local databases. National databases, such as NEI, can also be used when state and local data are not available. All stationary area sources will be modeled as point sources, except those shown on Table 11-1. It is envisioned that

only a few stationary area source emissions in a community will not be able to be described as discrete point sources. Table 11-1 shows the categories that need to be modeled as area sources, along with suggested allocation methods.

To model stationary area sources as a point source, the technical team will define a circle, centered on the centroid of each census tract. The area of the circle is equal to the area of the census tract. This is then repeated for each census tract located within the study area. This can be done using desktop mapping or GIS software (if available) and census tract data from the U.S. Census Bureau. The methodology described here uses pseudo-points but, if GIS resources are available, there is another perhaps easier method under development that uses GIS to define the boundaries of the census tract and model the emissions as an area source. Check the Community Assistance Technical Team web page for the availability of this methodology.

Five pseudo-stack source locations will be positioned inside each circle: one pseudo-stack source will be placed at the center of the circle (i.e., centroid of the census tract), and four pseudo-stack sources will be placed at a distance of one-half of the radius of the circle at each cardinal point (i.e., north, south, east, and west). The technical team will set the ISCST source-specific inputs to represent no plume rise, as below:

- Stack height: 5 meters
- Stack diameter: 0.001 meters
- Stack exit velocity: 0.001 meters/second
- Stack exit temperature: 298 Kelvin

The technical team will then weight the emissions from the pseudo-stack points accordingly: one-ninth of the emissions to the center pseudo-stack source and two-ninths of the emissions to each of the four surrounding pseudo-stack sources. Figure 11-2 depicts census tracts utilizing the pseudo-point source technique.

To allocate emissions, two different methods are available. One method is to estimate area source emissions using the 1996 NTI approach. A detailed discussion of the spatial allocation procedure appears in the NTI documentation on area sources (http://www.epa.gov/ttn/chief/emch/spatial/). Keep in mind that an inventory prepared by a state or local agency using county-specific data can include more local detail and assign emissions to the county level more accurately. The second method is to estimate emissions using the approach used for the Cumulative Exposure Project in 1998 (http://www.epa.gov/ttn/atw/urban/appx_b1.pdf). The methods each have advantages and disadvantages.

The advantage of the NTI method is simplicity. The method uses primarily population and employment information in its allocation methodology, but it is limited by the lack of employment at a smaller scale of resolution. This limitation increases uncertainty associated with area source emissions estimates. NTI assumes that the ratio between the total number of workers in an SIC code for the whole United States and the number of those workers in a given county is predictive of the ratio between national-scale emissions of chemicals in an SIC code and county emissions of individual chemicals in the SIC code. While this approach was deemed appropriate for allocation to the county level, the validity of the extension of this method to allocate from the county level to the census tract level has not been evaluated. This uncertainty is further reason why the Partnership, whenever possible, should attempt to determine the characteristics and locations of small and dispersed emissions and model them as point sources, avoiding most of the assumptions that they would have to make with area source emissions estimates.

In order to address the need to provide emissions at a finer level of resolution (e.g., census tract level), methods developed for this purpose by EPA in the Cumulative Exposure Project can also be used. The CEP work utilized an approach in which emissions were spatially allocated to the census tract level by using other geographically distributed surrogates including population and land use. See Appendix J for more information on the use of land use/land cover data and other methods of allocation.

Step 7: **Set up and run ISCST model to estimate total concentration from all sources.**

Once the information from steps 2 to 6 has been collected, it needs to be put into a data input file formatted for use during ISCST modeling runs. The input file format is very specific. The technical team should consult volume 1 of the ISCST user's guide (or commercial software support documents) to determine the specific data file format requirements. Figure 11-3 depicts a simplistic ISCST input file.

Each ISCST input file consists of six sections, or pathways:

- Control (CO) pathway
- Source (SO) pathway
- Receptor (RE) pathway
- Meteorological (ME) pathway
- Terrain grid (TG) pathway
- Output (OU) pathway

Below is a description of each pathway and some of the default assumptions and inputs that should be used when conducting modeling for the Secondary Screen.

Control Pathway
The Control pathway directs ISCST to perform specific types of computations. For the Secondary Screen, the technical team should use the following constraints: regulatory default options, concentrations, annual averaging times, and elevated terrain heights, if required. The technical team should also use the urban or rural land use setting defined for the study area.

Source Pathway
The Source pathway contains information on the source type and emission rate. Some key inputs in this pathway include:

- Location: developed in steps 5 and 6
- Emission rate: as described in the emissions inventory
- Source group: ALL
- Stack parameters for fugitive and area source releases

Building downwash effects will not be incorporated into the simplified version of the Secondary Screen used in this Manual.

Receptor Pathway
The Receptor pathway identifies the receptor grid nodes developed in step 1. These data are used by ISCST as the locations to compute estimates of air concentrations.

Meteorological Pathway
The Meteorological pathway directs ISCST to the input file containing the meteorological data. This pathway also defines the anemometer and mixing height information.

Table 11-1.
Area Source Categories and Apportioning Methods

SOURCE CATEGORY	NTI APPORTIONING METHOD	CEP APPORTIONING METHOD
Agricultural Production	Not specified	USGS land use categories: "farmland" and "orchard land," "confined feeding operations," "other agricultural land"
Asphalt Paving	County proportion of national 1996 population estimate [b]	Total miles of all roadway types in each census tract as reported in TIGER/Line
Consumer Products Usage 1996 population estimate [b]	County proportion of national	U.S. Census category. 1990 residential population
Gasoline Distribution Stage I	County proportion of national 1996 CBP employment for SIC code 5171	USGS land use categories: "industrial," plus one-half of "industrial and commercial complexes" plus one-third "mixed urban and built up land" plus one-third "other urban and built up land"
Industrial Boilers: Distillate Oil	County proportion of national 1996 CBP employment for SICGROUP=IND [a]	USGS land use categories: "commercial and services," plus one-half of "industrial and commercial complexes" plus one-third "mixed urban and built up land" plus one-third "other urban and built up land"
Institutional/Commercial Heating (all types)	County proportion of national 1996 CBP employment for SICGROUP=COMM [a]	USGS land use categories: "commercial and services," plus one-half of "industrial and commercial complexes" plus one-third "mixed urban and built up land" plus one-third "other urban and built up land"
Natural Gas Transmission and Storage	County proportion of national 1996 CBP employment for SIC code 4920: Gas Production and Distribution [a]	USGS land use categories: "industrial," plus one-half of "industrial and commercial complexes" plus one-third "mixed urban and built up land" plus one-third "other urban and built up land"
Non-industrial Asphalt Roofing	Not specified	Sum of USGS land use categories: "residential" and "commercial land"
Pesticide Application	County proportion of national 1990 PCBEIS agricultural acres (adjusted for 1996 county definitions) [c]	Not specified
Residential Heating (all types)	County proportion of national 1996 population estimate [b]	U.S. Census category. 1990 residential population

Table 11-1.
Area Source Categories and Apportioning Methods (continued)

SOURCE CATEGORY	NTI APPORTIONING METHOD	CEP APPORTIONING METHOD
Residential Heating, Wood	Regional proportion of national 1990 residential wood consumption and county proportion of regional 1996 population estimate [b]	Not specified
Structure Fires	County proportion of national 1996 population estimate [b]	U.S .Census category 1990 residential population
Surface Coatings: Architectural	County proportion of national 1996 population estimate [b]	USGS land use categories: "industrial," plus one-half of "industrial and commercial complexes" plus one-third "mixed urban and built up land" plus one-third "other urban and built up land"
Surface Coatings: Industrial Maintenance	County proportion of national 1996 CBP employment for SICGROUP=IND [a]	USGS land use categories: "industrial," plus one-half of "industrial and commercial complexes" plus one-third "mixed urban and built up land" plus one-third "other urban and built up land"
Surface Coatings: Traffic Markings	State proportion of 1996 national disbursements and county proportion of state 1996 population estimates [a]	USGS land use categories: "industrial," plus one-half of industrial and commercial complexes" plus one-third "mixed urban and built up land" plus one-third "other "urban and built up land"

a. *County Business Patterns 1996. United States Department of Commerce, Bureau of the Census. CBP-96-1. November 1998.*
b. *Estimates of the Population of Counties: Annual Time Series, July 1, 1990, to July 1, 1997 (includes revised April 1, 1990, census population counts). CO-97-4. Population Estimates Program, Population Division. U.S. Bureau of the Census, Washington, D.C.*
c. *The Personal Computer Biogenic Emissions Inventory System (PCBEIS) was developed by EPA and is available at http://www.epa.gov/ttn/chief/software.html.*
d. *Total Disbursements for Highways, All Units of Government - 1996. Table HF-2. United States Department of Transportation. October 1998.*

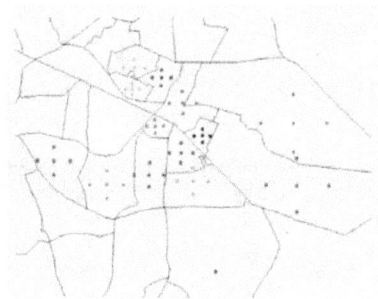

Figure 11-2.
Illustration of Pseudo-Point Layout

Figure 11-3.
Example of ISCST Input File Format

Terrain Grid Pathway

The Terrain Grid pathway defines input grid data used in calculating dry depletion in elevated or complex terrain. Depending on the decision of the technical team, dry depletion may or may not be considered.

Output Pathway

The Output pathway directs ISCST to generate outputs for reviewing, summarizing and plotting the air modeling results. The technical team should generate plot files depicting the annual average air concentrations for each emission source, at each receptor grid node.

Step 8: *Include background concentrations in model concentrations.*

Background concentrations are releases that are not the result of current human activity (including both natural and past human sources) in the study area and cannot be attributed to any of the stationary point, stationary area, or mobile sources that have been identified. During development of the Emission Source Inventory, background concentrations were identified for the various chemicals of concern. These values should be added to the concentrations developed from ISCST to estimate the total ambient concentration. The technical team should assume that the background concentrations collected in the Emission Source Inventory are uniform throughout the study area and are the same at each receptor location.

Step 9: *Compare modeled concentrations at monitoring site to available monitored concentrations.*

If the Partnership has access to relevant monitoring data for the study area, the technical team should use this data for comparison purposes with the ISCST model results for the study area monitoring sites. If the differences between the monitoring data and the modeled data are more than an order of magnitude, the technical team should review the values used during the modeling to identify the source of the discrepancies.

How will the technical team summarize its work to present to the full Partnership?

As mentioned at the beginning of this chapter, the results of the Secondary Screen will be used to identify those chemicals for further review in the Final Screen. As such, it is important for the technical team to provide results that support the Partnership's decision to retain or remove a chemical from consideration. A summary table should be compiled that lists the following for each of the analyzed chemicals:

- Name
- Maximum estimated concentration
- Screening-level concentration
- Ratio of maximum estimated concentration to screening-level concentration

The table should be arranged in descending order of the ratio of maximum chemical concentration to screening-level concentration. This type of table should provide the Partnership with an easy way to prioritize the chemicals for further analysis in the Final Screen.

The technical team should generate a summary table depicting the locations of sensitive populations selected for discrete concentration estimation. The table should be designed to display:

- Name of the location
- Name of the chemical
- Estimated concentration of the chemical
- Screening-level concentration
- Ratio of the estimated concentration to screening-level concentration

As with the first table, the chemicals in this table should be arranged in descending order of the ratio of estimated chemical concentration to screening-level concentration.

The technical team should also provide the Partnership with a table that lists the facilities releasing chemicals whose concentrations exceeded the screening level concentration. This table will provide the Partnership with a list of facilities to contact during the process for refinement of emission data in the Final Screen.

Finally, the technical team should provide the Partnership with a separate report detailing the technical team's air modeling approach, assumptions, data, and results to support the results depicted in the summary tables. If the Partnership has access to GIS resources, color-coded maps indicating the range of chemical concentrations throughout the study area can be developed.

What should the Partnership do after the ambient air concentration values have been estimated from each emission source?

Once the technical team is confident that there are no errors in the values they have estimated using the Secondary Screening approach, the Partnership needs to compare these estimates to the screening-level concentration values. Chemicals with concentrations above the screening level should be forwarded on to the Final Screen. Those chemicals with concentrations below the screening level can be set aside because they are below the community's screening levels.

Final Screen

What is the overall goal for the technical team?

The goal of the Final Screening process is to develop more accurate estimates of the ambient concentrations for chemicals remaining in the process after completion of the Secondary Screen, using the best information on sources that can be collected to develop new and more accurate estimates of concentrations in community air. With these new estimates, the Final Screen will be used to identify the chemicals and the sources that will be priorities for potential community action. A flow chart showing a summary of the steps for completing the Final Screen is shown in Figure 12-1.

How will the technical team estimate concentrations?

As it did in the Secondary Screen, the technical team will estimate chemical concentrations using the ISCST model. In addition, the MOBILE6.2 model may be used to develop more accurate concentration estimates for on-road sources.

How will the technical team collect the new information needed to estimate concentrations in the Final Screen?

The Partnership will draw on all of its members and organize special teams to contact, visit, or observe the sources that are under review in the Final Screen. This may involve surveying traffic on major roads or contacting and visiting both large and small commercial, industrial, and public facilities.

The Partnership teams that are established to collect the information should contain a cross section of Partnership members. Community leaders and local residents will explain the work of the Partnership and convince facility sources to cooperate and join in the effort. Technical members of the Partnership will be needed to help with the collection and verification of the information. Refer to the Final Screen chapter in the Overview for a description of the resources needed to collect the required information.

What new information will be used to estimate concentrations in the Final Screen?

The following is a list of the types of information the Partnership teams will have to collect for each source under review in the Final Screen.

Stationary Point Sources

Release amounts—Instead of using values obtained from various databases, the Partnership teams will collect and verify estimates or measurements of actual released amounts (e.g., tons/year). For those sources where a top-down or bottom-up approach was used to develop a stationary point source emission rate, the Partnership teams should visit the identified sites to see if the methods used to make the estimates were appropriate and investigate whether a better method could have been used.

Stack parameters—The Partnership teams will contact the facilities and collect and verify stack dimensions and locations. If a GPS is available, the Partnership teams may want to use it to verify emission locations. While at the facility, the Partnership teams should assess whether or not the buildings at the facility affect how the stack releases disperse into the surrounding area. If the buildings affect a chemical emission, downwash effects may need to be included in the ISCST modeling runs, and building dimensions will have to be collected.

Optional fugitive source dimensions for fugitive source modeling—The dimensions of areas at each facility where fugitive emissions occur if the technical team chooses to model fugitive emissions using the ISCST area source option.

Stationary Area Sources

The concentrations estimated for area sources in the Secondary Screen will likely be used again as the Final Screen estimations. The partnership should review the area source concentrations and determine if any further work is necessary to refine the estimated concentrations. This would involve revisiting the assumptions used for

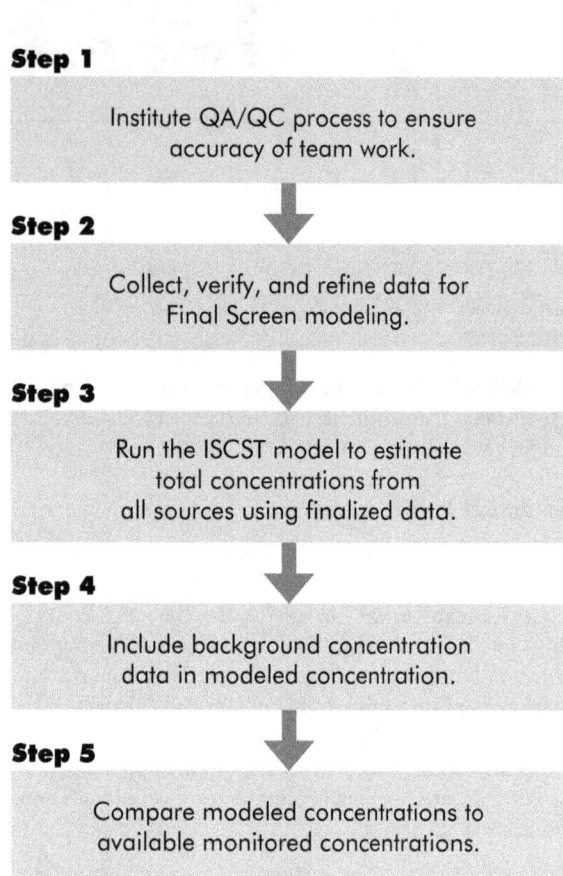

Step 1

Institute QA/QC process to ensure accuracy of team work.

Step 2

Collect, verify, and refine data for Final Screen modeling.

Step 3

Run the ISCST model to estimate total concentrations from all sources using finalized data.

Step 4

Include background concentration data in modeled concentration.

Step 5

Compare modeled concentrations to available monitored concentrations.

Step 6

USE ISCST model to estimate source contributions to total concentrations.

Figure 12-1.
Procedure for Final Screen

the pseudo-point source methodology during the Secondary Screen.

If the area source contributions are significant to the overall analysis, the Partnership may want to consider the use of an alternative to the pseudo-point modeling method used in the Secondary Screen. Stationary area source releases can be modeled as emissions released uniformly over the area of the census tract rather than combining the releases into five pseudo-stacks as done in the Secondary Screen. If GIS resources are available, this can be accomplished using the area source modeling option in ISCST and information on census tract boundaries. The documentation of the method is in preparation and should be available on this How-To Manual's wbsite at: http://www.epa.gov/oppt/cahp/howto.html.

Mobile Sources

The Partnership may decide that it has sufficient information to target mobile sources following the Secondary Screen. Refer to the discussion of mobile sources in the Final Screen chapter in the Overview. If the Partnership decides that more accurate estimations of mobile sources are necessary, the technical team will use the MOBILE6.2 model to estimate mobile soource emissions and ISCST air dispersion model to estimate concentrations. To run these models, the technical team will need to collect detailed information on traffic, fuels used, and the locations of main roadways and intersections in the Partnership area. Additional information on significant non-road sources may also be required. Examples can be found in the following references:

Stein, B. , Walker, D. Cook, R. and Bailey, C. 2003. *Link Based Calculation of Motor Vehicle Air Toxics Emissions Using MOBILE6.2.* EPA Office of Air Quality Planning and Standards. 12th International Emission Inventory Conference. April 29–May 1, 2003, San Diego, CA.

Thomas, G. and Dudley, M. 2001. *Steps in Conducting an Urban Air Toxics Assessment: Methodology for Converting Emissions Inventories into Model Ready Input Files.* USEPA Office of Air Quality Planning and Standards. 10th International Emission Inventory Conference. May 1 3, 2001, Denver, CO

Kinnee , EJ., Touma, JS, R. Mason, J. Thurmon, A. Beidler, C. Bailey and R. Cook. 2004. *Allocation of onroad mobile emissions to road segments for air toxics modeling in an urban area.* Transportation Research Part D: Transport and Environment. 9(2) 139-150.

Background Concentrations

Background concentrations are releases that are not the result of current human activity, including both natural and past human sources The background concentration of a chemical of interest, if available, will be used the same way it was in the Secondary Screen. It will be added to the modeled concentration estimates to get the total concentration of the chemical .

What are the steps the technical team will need to take to complete the work of estimating concentrations for the Final Screen?

Step 1: Institute QA/QC process to ensure accuracy of team work.

Step 2: Collect, verify, and refine data for Final Screen Modeling.

Step 3: Run the ISCST model using finalized data to estimate total concentrations from all sources.

Step 4: Add background concentrations to modeled concentrations.

Step 5: Compare modeled concentrations at monitoring site to available relevant monitored concentrations.

Step 6: Use the ISCST model to estimate source contributions to total concentrations for all chemicals with estimated concentrations greater than screening-level concentration.

How will the Partnership teams complete these steps?

Step 1: QA/QC

Refer to the QA/QC section (step 1) in Chapter 9 in the Technical Guidance section.

Step 2: Collect, verify, and refine data and methodologies used in Secondary Screen

The technical team will work with the teams organized by the Partnership to collect the detailed information that will be needed for the Final Screen. Teams will contact facility managers and visit stationary point sources to collect the information that will be needed for the Final Screen.

If the Partnership decides to refine its mobile source on-road estimations, a team will work with transportation department officials to collect the information needed to run the MOBILE6.2 and ISCST models. For detailed guidance on mobile source modeling using *MOBILE6.2, the Technical Description of the Toxics Module for MOBILE6.2 and Guidance on Its Use for Emission Inventory Preparation* (EPA 420-R-02-029, November 2002) can be found at http://www.epa.gov/otaq/models/mobile2/r02029.pdf.

If there is a significant non-road source or sources in the Partnership area, such as an airport, railroad yard, port, or heavy construction site with significant non-road emissions, the Partnership may decide to refine its non-road source estimations.

Step 3: Run the ISCST model using finalized data to estimate total concentrations from all sources.

Once the information from steps 1 and 2 has been collected, it needs to be put into a data input file for the ISCST and/or MOBILE6.2 modeling runs. As mentioned in the Secondary Screen, the input file format for modeling are very specific. The Partnership teams should consult the discussion about ISCST input files in the Secondary Screen section and volume 1 of the ISCST user's guide (or commercial software support documents) to determine the specific data file format requirements.

Step 4: Add background concentrations to modeled concentrations.

Background concentrations are levels of chemicals that occur naturally in the study area and cannot be attributed to any of the stationary point, stationary area, or mobile sources that have been identified. In the development of the Emission Source Inventory, background concentrations were identified for the various chemicals of concern. These values should be added to the concentrations developed from ISCST to estimate the total ambient concentration. The technical team should assume that the background concentrations collected in the emissions inventory are uniform throughout the study area and are the same at each receptor location.

Step 5: Compare modeled concentrations at monitoring sites to available monitored concentrations.

If the Partnership has access to monitoring data for the study area, the technical team should use this data for comparison purposes with the ISCST model results. If the differences between the monitoring data and the modeled data are more than an order of magnitude, the technical team should review the values used during the modeling to identify the source of the discrepancies.

Step 6: Use ISCST model to estimate source contributions to total concentrations.

The Partnership has identified a group of chemicals of concern and dropped other chemicals that are not of concern using the Secondary Screen. At this stage it is important to understand the contribution of each source of a chemical of concern to the concentration of that chemical from all sources at the receptor points chosen by the community. With this understanding it will be possible to identify candidate facilities for future voluntary emissions reduction efforts as well as the contribution of mobile and area sources to the concentrations of concern at the receptor points.

In order to determine individual source contributions to the total concentration of a chemical at a given receptor, multiple ISCST runs need to be made, as follows:

- Run ISCST for a single chemical and all sources of that chemical (stationary point, area, and mobile) using the best information available as collected in the Final Screen (e.g., measured or best estimated emissions, confirmed stack parameters and locations) and determine the chemical concentrations at the receptor locations of interest, including background concentration, if available.

- Edit input files or select options (as applicable) to model each source of the chemical one at a time and determine the concentration of the chemical at the receptor locations of interest. Be sure to treat area sources and mobile sources as groups of five pseudo-point sources in each census tract. When eliminating these sources to determine the contribution of a single point source of a chemical, all five pseudo-point sources in each census tract must be deleted or set to zero emissions.

- Determine the percentage of each source contribution to the total concentration at the receptor points of interest by dividing the concentration from the single source by the total

concentration from all sources determined in step 1. This can most easily be done by entering the data into a spreadsheet to carry out the calculations and save the information. In practice, it may not be necessary to account for 100% of the total concentration if it appears that there are only minor contributions from multiple facilities after greater than 95% of the emissions have been accounted for. The Partnership can make this decision if resources are limited.

When the results of this exercise are interpreted, more weight should be given to outcomes that indicate that stationary point sources are significant contributors than to those indicating that area sources are major contributors. For the stationary point sources, efforts have been made to refine the information on the emissions and stack characteristics as much as possible. Also, the results of ISCST modeling will be subject to less uncertainty than for area sources, where the uncertainty associated with emissions estimation, allocation, and the use of the pseudo-point source approach should be considered.

What is the next step after the technical team has estimated concentrations for the Final Screen?

Once the Final Screen concentrations have been estimated, the technical team should prepare a detailed report for the Partnership comparing estimated concentrations at each of the targeted community locations to the Partnership's screening-level concentrations. It may be helpful to include a ratio of the estimated concentration to the screening-level concentrations. For each chemical with a concentration above its screening level, the Final Screen summary report should also include a list of sources and their contribution, expressed as a percentage, to the total concentration. A graphical illustration of source contributions, such as a pie or bar chart, may be helpful. For the priority chemicals, it may also be helpful for the report to review and summarize all the information used to estimate concentrations and to develop the screening-level concentrations, describing any uncertainties and the overall level of confidence that the Partnership has in its analysis. The technical team will present this report to the full Partnership for discussion and approval and work with the full Partnership to help communicate the results of the screening to the community. This report will complete the technical work of the screening process. The priority chemicals and sources are now identified and the next steps for the Partnership, discussed in the final chapter of the Overview, will be to communicate the results to the community, develop recommendations for reducing exposures to the priority chemicals, and mobilize the community to implement the recommendations. Appendix G includes a list of resources for risk reduction and pollution prevention.

Glossary

Acute exposure: A single exposure to a toxic substance, which may result in severe biological harm or death. Acute exposures are usually characterized as lasting no longer than a day, as compared to longer, continuing exposure over a period of time.

Air contaminant: Any particulate matter, gas, or combination thereof, other than water vapor not normally found in air.

Air dispersion model: A computerized set of mathematical equations that uses emissions and meteorological information to simulate the behavior and movement of air pollutants in the atmosphere. The results of a dispersion model are estimated outdoor concentrations of individual air pollutants at specified locations. While air dispersion modeling relies on modeled estimates and not on actual measurements taken at monitoring stations, this method of estimating concentrations has been extensively tested and verified to provide reliable estimates of the concentrations that result from chemical releases.

Air pollutant: Any substance in air that could, in high enough concentration, harm man, other animals, vegetation, or material. Pollutants may include almost any natural or artificial composition of matter capable of being airborne. They may be in the form of solid particles, liquid droplets, gases, or in combination thereof. Generally, they fall into two main groups: (1) those emitted directly from identifiable sources, and (2) those produced in the air by interaction between two or more primary pollutants, or by reaction with normal atmospheric constituents, with or without photoactivation.

Air pollution: The presence of contaminants or pollutant substances in the air that interfere with human health or welfare, or produce other harmful environmental effects.

Air quality criteria: The levels of pollution and lengths of exposure above which adverse health and welfare effects may occur.

Air quality standards: The level of pollutants prescribed by regulations that are not to be exceeded during a given time in a defined area.

Air toxics: Also known as toxic air pollutants or hazardous air pollutants, those pollutants known to cause or suspected of causing cancer or other serious health problems. Health concerns may be associated with both short- and long-term exposures to these pollutants. Many are known to have respiratory, neurological, immune, or reproductive effects, particularly for more susceptible sensitive populations such as children.

Airborne particulates: Total suspended particulate matter found in the atmosphere as solid particles or liquid droplets. Chemical composition of particulates varies widely, depending on location and time of year. Sources of airborne particulates include dust, emissions from industrial processes, combustion products from the burning of wood and coal, combustion products associated with motor vehicle or non-road engine exhaust, and reactions to gases in the atmosphere.

Ambient: Surrounding, as in the surrounding environment. In this assessment, ambient air refers to the air surrounding a person through which pollutants can be carried.

Ambient air: Any unconfined portion of the atmosphere; open air, surrounding air.

Ambient air monitoring station: Ambient air monitoring stations collect air samples on a regular basis to determine the concentration of chemicals present in the air. Monitored concentrations are an accurate measure of concentrations at specific locations.

Background concentration: In this Manual, the contributions to outdoor air toxics concentrations resulting from natural sources, persistence in the environment of emissions from previous years, and long-range transport from distant sources. Background concentrations could be levels of pollutants that would be found even if there had been no recent man-made emissions. To accurately estimate outdoor concentrations, it is necessary to account for the background concentrations by adding them to the modeled concentrations. The National Air Toxics Assessment (NATA) program provides background concentrations based on monitored values identified in the Cumulative Exposure Project (CEP). The CEP is a study that estimated nationwide 1990 ambient concentrations of air toxics. Based on that study, nationwide background concentration values for 13 toxic air pollutants were developed. Details for including these background concentrations in the Emission Source Inventory can be found in Chapter 9, "Building the Emission Source Inventory."

Carcinogen: A chemical or physical agent capable of causing cancer.

CAS registration number: A number assigned by the Chemical Abstract Service to identify a chemical.

Census tracts: Land areas defined by the U.S. Bureau of the Census that vary in size but typically contain about 4,000 residents each. Census tracts are usually smaller than two square miles in cities but are much larger in rural areas.

Chronic effect: An adverse effect on a human or animal in which symptoms recur frequently or develop slowly over a long period of time.

Chronic exposure: Multiple exposures occurring over an extended period of time or over a significant fraction of an animal's or human's lifetime (usually seven years to a lifetime).

Chronic toxicity: The capacity of a substance to cause long-term poisonous health effects in humans, animals, fish, and other organisms.

Compliance monitoring: Collection and evaluation of data, including self-monitoring reports, and verification to show whether pollutant concentrations and loads contained in permitted discharges are in compliance with the limits and conditions specified in the permit.

Concentration: The relative amount of a substance mixed with another substance. An example is 5 ppm of carbon monoxide in air or 1 mg/L of iron in water.

Constituents of concern: Specific chemicals that are identified for evaluation in the site assessment process.

Contaminant: Any physical, chemical, biological, or radiological substance or matter that has an adverse effect on air, water, or soil.

Criteria pollutants: Following the 1990 amendments to the Clean Air Act, a term derived from the requirement that EPA must describe the characteristics and potential health and welfare effects of these pollutants. The amendments required EPA to set National Ambient Air Quality Standards for certain pollutants known to be hazardous to human health. EPA has identified and set standards to protect human health and welfare for six pollutants: ozone, carbon monoxide, total suspended particulates, sulfur dioxide, lead, and nitrogen oxide. It is on the basis of these criteria that standards are set or revised.

Cubic feet per minute (CFM): A measure of the volume of a substance flowing through air within a fixed period of time. With regard to indoor air, refers to the amount of air, in cubic feet, that is exchanged with outdoor air in one minute, i.e., the air exchange rate.

Cumulative exposure: The sum of exposures of an organism to a pollutant over a period of time.

Cumulative risk: An analysis, characterization, and possible quantification of the combined risks to health or the environment from multiple agents or stressors.

Designated pollutant: An air pollutant that is neither a criteria nor hazardous pollutant, as described in the Clean Air Act, but for which new source performance standards exist. The Clean Air Act does require states to control these pollutants, which include acid mist, total reduced sulfur (TRS), and fluorides.

Diesel particulate matter: A mixture of particles that is a component of diesel exhaust. EPA lists diesel exhaust as a mobile source air toxic due to the cancer and non-cancer health effects associated with exposure to whole diesel exhaust. EPA believes that exposure to whole diesel exhaust is best described, as many researchers have done over the years, by diesel particulate concentrations.

Dispersion model: See Air Dispersion Model

Emission Source Inventory: A collection of information on all the known sources of air emissions in and around the Partnership area. The Emission Source Inventory contains all the information on sources that the Partnership will need to complete all the steps of the air screening exercise described in this Manual. The inventory will serve as a permanent community database that can be updated periodically and used to measure progress in improving community air quality.

Emission standard: The maximum amount of air polluting discharge legally allowed from a single source, mobile or stationary.

Exposure assessment: Identifcation of the ways in which chemicals may reach individuals (e.g., by breathing); estimation of how much of a chemical an individual is likely to be exposed to; and estimation of the number of individuals likely to be exposed.

Fugitive source: Stationary point source releases are divided into stack and fugitive release types in this Manual. Examples of stack releases include a release from a roof vent of a dry-cleaning business and a release from a large stack like the kind used for most municipal trash incinerators. Examples of fugitive releases include emissions from the surface of a treatment pond or from the nozzles at a gas filling station.

Geographic Information System (GIS): Can be used to illustrate the sources and their locations in and around the Partnership area.

Hazard quotient: The ratio of the potential exposure to the substance and the level at which no adverse effects are expected. If the hazard quotient is calculated to be less than 1, then no adverse health effects are expected as a result of exposure. If the hazard quotient is greater than 1, then adverse health effects are possible. The hazard quotient cannot be translated to a probability that adverse health effects will occur, and it is unlikely to be proportional to risk. It is especially important to note that a hazard quotient exceeding 1 does not necessarily mean that adverse effects will occur.

Hazardous Air Pollutant Exposure Model, Version 4 (HAPEM4): A computer model that has been designed to estimate inhalation exposure for specified population groups and air toxics. Through a series of calculation routines, the model makes use of census data, human activity patterns, ambient air quality levels, climate data, and indoor/outdoor concentration relationships to estimate an expected range of inhalation exposure concentrations for groups of individuals.

Hazardous air pollutants (HAPs): The 188 air pollutants that are not covered by ambient air quality standards but that, as defined in the Clean Air Act, may present a threat of adverse human health effects or adverse environmental effects. Such pollutants include asbestos, beryllium, benzene, coke oven emissions, mercury, radionuclides, and vinyl chloride. The full list of HAPs is given in Appendix E.

Heavy metals: Metallic elements with high atomic weights (e.g., mercury, chromium, cadmium, arsenic, and lead) that can damage living things at low concentrations and tend to accumulate in the food chain.

Human health risk: The likelihood that a given exposure or series of exposures may have damaged or will damage the health of individuals.

Inhalation: Breathing. Once inhaled, contaminants can be deposited in the lungs, taken into the blood, or both.

Inhalation reference dose: An estimate (with uncertainty spanning perhaps an order of magnitude) of daily level of inhalation exposure of the human population (including sensitive subgroups) that is likely to be without an appreciable risk of adverse effects during a lifetime.

IRIS: EPA's Integrated Risk Information System, an electronic database containing the Agency's latest descriptive and quantitative regulatory information on chemical constituents.

Industrial Source Complex Short Term (ISCST) air dispersion model: A widely used model developed by EPA to estimate the air concentrations that result from chemical releases.

Local knowledge: The knowledge that members of the community, including residents and local businesses, have as a result of living and working in the community. The local knowledge important to risk-based screening includes detailed information on local sources and releases, on ways that community members are exposed to toxics, and on possible solutions to address the priorities identified by the screening exercise. Local knowledge also includes a detailed understanding of the community values, organization, and resources that will be needed to effectively mobilize the community to address local air quality concerns.

Microgram: One millionth of a gram. One gram is about 1/28th of an ounce.

Mobile source: Any non-stationary source of air pollution such as cars, trucks, motorcycles, buses, airplanes, and locomotives.

Monitoring: Periodic or continuous surveillance or testing to determine the level of compliance with statutory requirements and/or pollutant levels in various media or in humans, plants, and animals. Monitoring directly measures what is in the air, either at fixed locations in a community or as the air is breathed by someone living in the community through personal monitors worn by community volunteers.

National Air Toxics Assessment (NATA): EPA's ongoing comprehensive evaluation of air toxics in the United States. These activities include expanding air toxics monitoring, improving and periodically updating emission inventories, improving national- and local-scale modeling, researching health effects and exposures to both ambient and indoor air, and improving assessment tools.

National Ambient Air Quality Standards (NAAQS): Standards established by EPA that apply for outdoor air throughout the country.

National Emissions Inventory (NEI): EPA's compilation of quantitative information concerning the mass of air toxics emitted into the atmosphere (through smokestacks, tailpipes, vents, etc.). The NEI provides a model-ready emissions inventory as part of the National Air Toxics Assessment required by the Clean Air Act.

Mobile sources—non-road: Mobile sources not found on roads and highways (e.g., airplanes, trains, lawn mowers, construction vehicles, farm machinery).

Mobile sources—on-road: Vehicles found on roads and highways (e.g., cars, trucks, buses).

One in 1 million cancer risk: A risk level of one in 1 million implies a likelihood that up to one person, out of 1 million equally exposed people, would contract cancer if exposed continuously (24 hours per day) to the specific concentration over 70 years (an assumed lifetime). This would be in addition to those cancer cases that would normally occur in an unexposed population of 1 million people. Note that this assessment looks at lifetime cancer risks, which should not be confused with or compared to annual cancer risk estimates. If you would like to compare an annual cancer risk estimate with the results in this assessment, you would need to multiply that annual estimate by a factor of 70 or alternatively divide the lifetime risk by a factor of 70. A one in 1 million lifetime risk to the public in 1996 was 250 cancer cases over a 70-year period.

Organic chemicals/compounds: Naturally occurring (animal or plant produced) or synthetic substances containing mainly carbon, hydrogen, nitrogen, and oxygen.

Order of magnitude: A factor of 10 difference between two values (e.g., the difference between 0.1 and 1, or 10 and 100 is an order of magnitude).

Peak levels: Levels of airborne pollutant contaminants much higher than average or occurring for short periods of time in response to sudden releases.

Plume: A visible or measurable discharge of a contaminant from a given point of origin, such as a plume of smoke or the area downwind from a release, where exposures to the releases can occur and concentrations can be estimated.

PM 10, PM 2.5: PM 10 is measure of particles in the atmosphere with a diameter of less than 10 or equal to a nominal 10 micrometers. PM 2.5 is a measure of smaller particles in the air. PM 10 has been the pollutant particulate-level standard against which EPA has been measuring Clean Air Act compliance. On the basis of newer scientific findings, the Agency is considering regulations that will make PM 2.5 the new standard.

Pollutant: Generally, any substance introduced into the environment that adversely affects the usefulness of a resource or the health of humans, animals, or ecosystems..

Population at risk: A population subgroup that is more likely to be exposed to a chemical, or is more sensitive to the chemical, than is the general population.

Pseudo point: The artificial point sources given specific locations that are used in air dispersion modeling to represent sources of pollution that are too numerous and disperse to model individually (e.g., home heating, consumer product use, lawn mower exhaust, etc.). Pseudo points only exist in the model and they do not have an actual location in the real world. Pseudo points are used as a means to estimate the contribution of dispersed emissions to airborne concentrations of pollutants. For example, in the modeling recommendations included in this Manual, to make air dispersion modeling practical, all the emissions from multiple home heating systems in a census tract are combined and modeled as though they are released from five pseudo points spaced evenly throughout the census tract.

Quality assurance/quality control (QA/QC): A system of procedures, checks, audits, and corrective actions to ensure that all EPA research design and performance, environmental monitoring and sampling, and other technical and reporting activities are of the highest achievable quality.

Reference concentration (RfC): The reference

concentration is an estimate (with uncertainty spanning perhaps an order of magnitude) of a continuous inhalation exposure to the human population (including sensitive subgroups which include children, asthmatics and the elderly) that is likely to be without an appreciable risk of deleterious effects during a lifetime. It can be derived from various types of human or animal data, with uncertainty factors generally applied to reflect limitations of the data used.

Risk: The probability that damage to life, health, and/or the environment will occur as a result of a given hazard (such as exposure to a toxic chemical). Some risks can be measured or estimated in numerical terms (e.g., one chance in a million).

Risk characterization: The last phase of the risk assessment process that estimates the potential for adverse health or ecological effects to occur from exposure to a stressor and evaluates the uncertainty involved.

Risk communication: The exchange of information about health or environmental risks among risk assessors and managers, the general public, news media, interest groups, etc.

Route of exposure: The avenue by which a chemical comes into contact with an organism, e.g., inhalation, ingestion, dermal contact, injection.

Risk-based screening: Risk-based screening is designed to identify the chemicals and sources that present the greatest risk to the community with the minimum possible expenditure of community resources. Risk based screening minimizes resources by screening out low risk chemicals and simplifying the estimation of exposure to develop relative risk estimates for each chemical concentration and each source. Because of these simplifications, risk based screening has limitations. It can identify chemicals above screening levels, rank chemicals and sources by risk, and identify the chemicals and sources with the highest risk, but risk based screening does not estimate the risks resulting from releases and sources. And, because risk based screening does not attempt to estimate risk, it can not be used to estimate the cumulative risk from all or some of the chemicals.

Risk screening level: The risk screening level is a level of risk that the community agrees to use to identify priorities. This risk level is used to sort through the concentrations in community air to find those concentrations that pose a potential risk greater than the

risk screening level. These concentrations and the sources that contribute to them become the priorities that will be evaluated for possible community action.

Screening-level concentration: The screening-level concentration is the air concentration that would result after a lifetime of exposure in a risk equal to the screening level of risk chosen by the Partnership. Once the Partnership has chosen a risk screening level to use for screening, it will next need to calculate, for each chemical in community air, the air concentration that corresponds to the community's risk screening level. The concentrations in community air estimated for the Initial Screen will be compared to the screening-level concentrations.

Stack: A chimney, smokestack, or vertical pipe that discharges used air.

State, tribal, and local government air permitting databases: The primary source for government information on local air sources. State, tribal, or local government authorities are responsible for permitting air releases. They maintain databases with information on local sources and their releases. The amount and kind of information collected will depend on the local authority.

Stationary sources: Stationary sources include all the emission sources that come from fixed locations. This type of source includes both large and small stationary sources ranging from large sources such as electric utility plants, chemical plants, steel mills, oil refineries, and hazardous waste incinerators to small stationary sources such as the neighborhood dry-cleaners and auto repair and refinishing shops. Small stationary sources also include all the emission sources from homes and small office buildings, such as wood stoves, gas-fired home heating, and even household chemical uses.

Stationary area sources: As defined in this Manual, area sources, also called stationary area sources, include sources, such as home furnaces and wood stoves that are too small and ubiquitous in nature to be inventoried as individual sources. Because these sources are too numerous to estimate air concentrations individually, the releases from these sources will be combined together by type. The Manual defines the list of sources that will be handled as area sources for risk screening. To avoid confusion, it is important to note that the definition of area source used in this Manual differs from the definition used in EPA's Air Program. Area sources as defined in the Air Program can include small stationary sources such as dry-cleaners and gas stations. In this Manual, all small commercial and business sources will be handled as stationary point sources, and the concentrations resulting from the releases from these sources will be estimated individually.

Stationary point sources: Include all the stationary sources whose concentrations will be estimated individually. Large and small businesses, from large chemical facilities and steel mills to the gasoline station on the neighborhood corner, will be included in this group. Publicly owned facilities such as incinerators and water treatment plants will also be put into this group. Each of these facilities will have the concentrations resulting from its releases estimated individually.

Toxics Release Inventory (TRI): This EPA database contains information on air releases of over 600 chemicals from the large facilities across the country.

Toxic substance: A chemical or mixture that can cause illness, death, disease, or birth defects. The quantities and exposures necessary to cause these effects can vary widely. Many toxic substances are pollutants and contaminants in the environment.

Upper bound: A plausible upper limit to the true value of a quantity; usually not a true statistical confidence limit.

Upper-bound lifetime cancer risk: A plausible upper limit to the true probability that an individual will contract cancer over a 70-year lifetime as a result of a given hazard (such as exposure to a toxic chemical). This risk can be measured or estimated in numerical terms (e.g., one chance in a hundred).

Unit risk estimate: The upper-bound excess lifetime cancer risk estimated to result from continuous exposure to an agent at a concentration of 1 $\mu g/m^3$ in air. The interpretation of the unit risk estimate would be as follows: If the unit risk estimate equals 1.5×10^6 per $\mu g/m^3$, 1.5 excess tumors are expected to develop per 1,000,000 people exposed daily for a lifetime to 1 μg of the chemical in 1 cubic meter of air. Unit risk estimates are considered upper-bound estimates, meaning they represent a plausible upper limit to the true value. (Note that this is usually not a true statistical confidence limit.) The true risk is likely to be less, but could be greater.

Vehicle miles traveled (VMT): A measure of the extent of motor vehicle operation; the total number of vehicle miles traveled within a specific geographic area over a given period of time.

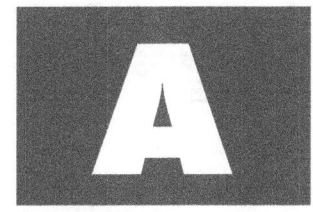

Appendix

Resources for Building a Partnership, Setting Goals, and Developing a Communication Plan

The links below provide access to both web-based and printed resources for building partnerships, setting goals, and developing a plan for communication. The list of web sites is a only a sample of the resources that are available on the Internet. Each of these sources provides links to additional resources that are available. Please check the Manual web site for updates to these links.

In addition to resources available on the Internet, local libraries and librarians are also an excellent source for information on these topics. Local community organizations and colleges can also be great resources to help you get started and build your organization.

EPA and other federal government resources

Community-Based Environmental Protection
See the tool, resources, and links pages on this site:
http://www.epa.gov/ecocommunity/about.htm

Watershed Information Network
Although focused on watersheds, this site has excellent resources for getting started with a community project:
http://www.epa.gov/win/ and
http://www.epa.gov/win/start.html

Community Involvement Toolkit
This excellent community resource was developed for the Superfund program:
http://www.epa.gov/superfund/tools/index.htm

Public Involvement site
For resources see the tools page on this site:
http://www.epa.gov/stakeholders/

Department of Housing and Urban Development (HUD), Communities site
http://www.hud.gov/community/index.cfm

National Park Service Community Toolbox
http://www.nps.gov/phso/rtcatoolbox/

Other resources

Please note that the following links are not a part of the EPA.gov domain. These links provide additional information that may be useful or interesting and are being provided consistent with the intended purpose of this EPA document. However, EPA cannot attest to the accuracy of information provided by these links or any other linked site. Providing links to a non-EPA web site does not constitute an endorsement by EPA or any of its employees of the sponsors of the site or the information or products presented on the site. Also, be aware that the privacy protection provided on the epa.gov domain may not be available at the external link.

Asset Based Community Development Institute
http://www.northwestern.edu/ipr/abcd.html

Center for Collaborative Planning, Resource Library
http://www.connectccp.org/resources/

Civic Practices Network Tools Community
http://www.cpn.org/tools/manuals/community/index.html

Citizens Handbook, A Guide to Building Community
http://www.vcn.bc.ca/citizens handbook/

Clean Air Counts, Campaigns for Clean Air and Development, Metropolitan Chicago
http://www.cleanaircounts.org/

Community Campus Partnerships for Health
http://www.futurehealth.ucsf.edu/ccph.html

Community Problem Solving, Strategy for a Changing World
http://www.community problem solving.net/

Community Tool Box
http://ctb.ku.edu/

National Civic League
http://www.ncl.org/

Community Visioning and Strategic Planning Handbook
http://www.ncl.org/publications/online/VSPHandbook.pdf

Partnership for Public Health, Resources
http://www.partnershipph.org/col3/resc/res_index.html

Study Circle Resource Center
http://www.studycircles.org/index.html

Sustainable Communities Network
http://www.sustainable.org/

Appendix

Examples of Projects Communities Have Adopted to Improve Air Quality

Examples of the risk reduction activities or options that have been identified for consideration by community partnerships in Cleveland, Ohio; St. Louis, Missouri; and West Oakland, California, can be found at the web sites provided below. The web site for the Clean Air Counts program from the metropolitan Chicago area is also provided as a resource for risk reduction activities.

In addition, for a list of activities that communities are taking to reduce exposure to diesel exhaust from school buses see the descriptions of the Clean School Bus Demonstration Projects at http://www.epa.gov/otaq/schoolbus/grants_2003.htm.

The web sites listed below do not provide a comprehensive list of activities that communities have used to reduce exposures and risks. At the time of the publication of this Manual, many communities were completing local assessments and beginning to consider options for reductions, so many more examples of reduction activities will be available in the future. As communities complete their reduction strategies, descriptions of their plans will be made available on EPA's **Air Toxics Community Assessment and Risk Reduction Projects Database** at http://yosemite.epa.gov/oar/CommunityAssessment.nsf/Welcome?OpenForm.

For a list of resources and programs for pollution prevention and emissions reduction please see Appendix G.

Please note that the following links are not a part of the EPA.gov domain. These links provide additional information that may be useful or interesting and are being provided consistent with the intended purpose of this EPA document. However, EPA cannot attest to the accuracy of information provided by these links or any other linked site. Providing links to a non-EPA web site does not constitute an endorsement by EPA or any of its employees of the sponsors of the site or the information or products presented on the site. Also, be aware that the privacy protection provided on the epa.gov domain may not be available at the external link.

Clean Air Counts, Campaigns for Clean Air and Development, Metropolitan Chicago
http://www.cleanaircounts.org/

Cleveland Air Toxics Pilot Project
http://www.ohiolung.org/ccacc.htm

St. Louis Community Air Project
http://www.stlcap.org/whatsnew.htm

One of the results of the St. Louis work is a set of curriculum materials developed by the Missouri Botanical Gardens. These educational materials are designed to help learners of all ages understand air toxics and how they can affect health and what can be done to reduce air toxics levels. These materials are developed in five units for kindergarten through adult learners. These curriculum materials will be available in the spring of 2004 at http://www.mobot.org/gatewaycenter/.

West Oakland: Report on Reducing Diesel Pollution
http://www.pacinst.org/diesel/index.html

Appendix

Resources for Addressing Indoor Air Exposures and Acute Outdoor Air Exposures

The following list of resources is not comprehensive. Please use these resources as a starting point for your work. Please see Appendix G for risk reduction resources for indoor air.

Please note that some of the following links are not a part of the epa.gov domain. These links provide additional information that may be useful or interesting and are being provided consistent with the intended purpose of this EPA document. However, EPA cannot attest to the accuracy of information provided by these links or any other linked site. Providing links to a non-EPA web site does not constitute an endorsement by EPA or any of its employees of the sponsors of the site or the information or products presented on the site. Also, be aware that the privacy protection provided on the epa.gov domain may not be available at the external link.

Indoor Air Resources

Indoor Air Quality Home Page
http://www.epa.gov/iaq/

An Introduction to Indoor Air Quality (IAQ)
http://www.epa.gov/iaq/ia-intro.html

The Inside Story: A Guide to Indoor Air Quality
http://www.epa.gov/iaq/pubs/insidest.html

Fact Sheet: Respiratory Health Effects of Passive Smoking
http://www.epa.gov/smokefree/pubs/etsfs.html

American Lung Association
http://www.lungusa.org/

Community Environmental Health Resource Center (CEHRC)
http://www.cehrc.org/

California Indoor Air Quality Program
http://www.cal-iaq.org/

Indoor Air Quality Association
http://www.iaqa.org/

Sources for Information on Acute Effects of Toxic Chemicals

Agency for Toxic Substance and Disease Registry Minimal Risk Levels for Hazardous Substances
http://www.atsdr.cdc.gov/mrls.html

California Environmental Protection Agency Office of Environmental Health Hazard Assessment Toxicity Criteria Database
http://www.oehha.ca.gov/risk/chemicalDB/index.asp

U.S. EPA Integrated Risk Information System (IRIS) 2003
http://www.epa.gov/ngispgm3/iris/subst/index.html

U.S. EPA Office of Air Quality Planning and Standards Air Toxics Dose-Response Database 2002
http://www.epa.gov/ttn/atw/toxsource/summary120202.html

Technical Background Document to Support Rulemaking Pursuant to the Clean Air Act - Section 112 (g): Ranking of Pollutants with Respect to Hazard to Human Health, U.S. Environmental Protection Agency. 1994. EPA-450/3-92-010, February 1994.

Voluntary Remediation Program (VRP) Risk Assessment Guidance, Virginia Department of Environmental Quality (VDEQ). 2000.

Summary of Lessons Learned by the Baltimore Partnership

Lessons Learned in Baltimore

The work of the Community Environmental Partnership in Baltimore, Maryland, was a learning experience for all of the people who participated. The Partnership tried a lot of new things—some of them worked and some didn't. A few lessons learned from this work are listed below. A complete list of lessons learned by the Baltimore Air Study can be found in the case study at http://www.epa.gov.opptintr/cahp/case.html.

The following are among the lessons learned:

- Bringing community stakeholders and science experts together to address community concerns is a key to success. Both the local knowledge of community residents and the science and tools of experts are needed to understand and to find the most effective ways to improve a local environment. Developing a dialogue between residents and experts also helps to build the consensus that will be needed to take action.

- Don't skimp on the time spent at the beginning of a project clarifying goals. Make sure all the participants are clear about their own goals and communicate them to the other members of the partnership. Make sure participants are clear about how the goals set by the partnership relate to their own goals. If the goals of the project match the participants' goals, the resources and energy will be sufficient to sustain the work.

- Building a strong partnership with a full range of stakeholders and broad community participation is key to mobilizing the new resources that will be needed for detailed local assessments and for implementing local solutions. No single community group or level of government has the resources to address local issues, so building a partnership and learning to work together will be essential. Local community organizations and local government can form the core of the partnership, with others outside the community providing resources not available at the local level.

- Plan for adequate time and resources to build and sustain the partnership. Working in a broad partnership using science at the community level will be a new way of doing business for most of the participants. Take time to build the trust, develop the organization, and provide the information and training that all the participants will need to work effectively in a partnership.

- Once goals for the partnership are clarified, make sure the partnership has the resources and scope to meet the goals. For example, if improving the health of the community is a goal, a broad partnership that can address all of the important factors affecting community health will need to be formed. Any attempt to improve community health by addressing environmental issues by themselves will probably not be able to produce results and will lead to disappointment.

- Be prepared to take a step-by-step approach to building the consensus on environmental issues in your community. Communities may decide to focus on their main concerns first and take on other areas in the future. It will take time to complete a fuller picture of the local environment that everyone can agree with.

- Be prepared to be creative in solving problems. The information, analysis tools, and solutions that communities need to improve their environments do not come ready-made. Your partnership will probably have to find ways to collect data and use and adapt tools designed for other purposes to answer your questions. If it is broad enough, your partnership will have the resources to find answers, but only if it is prepared to be creative. Making progress at the local level will depend on effective sharing among communities, so when you develop new approaches, please share them so that other communities do not have to reinvent the wheel. If you have ideas you can share with others, send them to us and we will post them on this Manual's web site for others to learn from.

- Communicate and involve the broader community in the work of the partnership on a regular basis. Regular communication gives the broader community an opportunity to provide input into the work of the partnership, participate in partnership activities, and learn from the work. Identifying community needs, taking advantage of community knowledge, and mobilizing the community to take action all will depend on maintaining a close and active relationship between the partnership organization and the community.

- Pay attention to the long-term capacity of the community to address environmental concerns. Identify areas of community capacity that need strengthening and organize the work of the partnership to help build capacity in these areas. In addition to knowledge and training in environmental areas, include other issues that will be important to long-term capacity, such as fund-raising, organizing, and leadership skills.

- Recognize that the work to understand and improve local environments will take a sustained long-term effort, and getting to measurable results will take years. Partnerships should take a long-term perspective and develop plans accordingly. The contributions of outside partners may vary according to their priorities, but those contributions should be adjusted to fit into the Partnership's plan for sustaining a long-term effort.

List of Hazardous Air Pollutants (HAPs)

CAS Number	Chemical	CAS Number	Chemical
75070	Acetaldehyde	542881	Bis(chloromethyl)ether
60355	Acetamide	75252	Bromoform
75058	Acetonitrile	106990	1,3-Butadiene
98862	Acetophenone	156627	Calcium cyanamide
53963	2-Acetylaminofluorene	105602	Caprolactam (See Modification)
107028	Acrolein		
79061	Acrylamide	133062	Captan
79107	Acrylic acid	63252	Carbaryl
107131	Acrylonitrile	75150	Carbon disulfide
107051	Allyl chloride	56235	Carbon tetrachloride
92671	4-Aminobiphenyl	463581	Carbonyl sulfide
62533	Aniline	120809	Catechol
90040	o-Anisidine	133904	Chloramben
1332214	Asbestos	57749	Chlordane
71432	Benzene (including benzene from gasoline)	7782505	Chlorine
		79118	Chloroacetic acid
92875	Benzidine	532274	2-Chloroacetophenone
98077	Benzotrichloride	108907	Chlorobenzene
100447	Benzyl chloride	510156	Chlorobenzilate
92524	Biphenyl	67663	Chloroform
117817	Bis(2-ethylhexyl)phthalate (DEHP)	107302	Chloromethyl methyl ether

The original list of Hazardous Air Pollutants can also be found at the EPA Technology Transfer Network Air Toxics web site:
http//www.epa.gov/ttn/atw/188pols.html

Modifications to the 112(b)1 Hazardous Air Pollutants can also be found at the EPA Technology Transfer Network Air Toxics web site:
http//www.epa.gov/ttn/atwsmod/188pols.html

CAS Number	Chemical
126998	Chloroprene
1319773	Cresols/Cresylic acid (isomers and mixture)
108394	m-Cresol
95487	o-Cresol
106445	p-Cresol
98828	Cumene
94757	2,4-D, salts and esters
3547044	DDE
334883	Diazomethane
132649	Dibenzofurans
96128	1,2-Dibromo-3-chloropropane
84742	Dibutylphthalate
106467	1,4-Dichlorobenzene(p)
91941	3,3-Dichlorobenzidene
111444	Dichloroethyl ether (Bis(2-chloroethyl)ether)
542756	1,3-Dichloropropene
62737	Dichlorvos
111422	Diethanolamine
121697	N,N-Diethyl aniline (N,N-Dimethylaniline)
64675	Diethyl sulfate
119904	3,3-Dimethoxybenzidine
60117	Dimethyl aminoazobenzene
119937	3,3'-Dimethyl benzidine
79447	Dimethyl carbamoyl chloride
68122	Dimethyl formamide
57147	1,1-Dimethyl hydrazine
131113	Dimethyl phthalate
77781	Dimethyl sulfate
534521	4,6-Dinitro-o-cresol, and salts
51285	2,4-Dinitrophenol

CAS Number	Chemical
121142	2,4-Dinitrotoluene
123911	1,4-Dioxane (1,4-Diethyleneoxide)
122667	1,2-Diphenylhydrazine
106898	Epichlorohydrin (l-Chloro-2,3-epoxypropane)
106887	1,2-Epoxybutane
140885	Ethyl acrylate
100414	Ethyl benzene
51796	Ethyl carbamate (Urethane)
75003	Ethyl chloride (Chloroethane)
106934	Ethylene dibromide (Dibromoethane)
107062	Ethylene dichloride (1,2-Dichloroethane)
107211	Ethylene glycol
151564	Ethylene imine (Aziridine)
75218	Ethylene oxide
96457	Ethylene thiourea
75343	Ethylidene dichloride (1,1-Dichloroethane)
50000	Formaldehyde
76448	Heptachlor
118741	Hexachlorobenzene
87683	Hexachlorobutadiene
77474	Hexachlorocyclopentadiene
67721	Hexachloroethane
822060	Hexamethylene-1,6-diisocyanate
680319	Hexamethylphosphoramide
110543	Hexane
302012	Hydrazine
7647010	Hydrochloric acid

CAS Number	Chemical	CAS Number	Chemical
7664393	Hydrogen fluoride (Hydrofluoric acid)	92933	4-Nitrobiphenyl
		100027	4-Nitrophenol
7783064	Hydrogen sulfide (See Modification)	79469	2-Nitropropane
		684935	N-Nitroso-N-methylurea
123319	Hydroquinone	62759	N-Nitrosodimethylamine
78591	Isophorone	59892	N-Nitrosomorpholine
58899	Lindane (all isomers)	56382	Parathion
108316	Maleic anhydride	82688	Pentachloronitrobenzene (Quintobenzene)
67561	Methanol	87865	Pentachlorophenol
72435	Methoxychlor	108952	Phenol
74839	Methyl bromide (Bromomethane)	106503	p-Phenylenediamine
74873	Methyl chloride (Chloromethane)	75445	Phosgene
		7803512	Phosphine
71556	Methyl chloroform (1,1,1-Trichloroethane)	7723140	Phosphorus
		85449	Phthalic anhydride
78933	Methyl ethyl ketone (2-Butanone)	1336363	Polychlorinated biphenyls (Aroclors)
60344	Methyl hydrazine	1120714	1,3-Propane sultone
74884	Methyl iodide (Iodomethane)	57578	beta-Propiolactone
108101	Methyl isobutyl ketone (Hexone)	123386	Propionaldehyde
		114261	Propoxur (Baygon)
624839	Methyl isocyanate	78875	Propylene dichloride (1,2-Dichloropropane)
80626	Methyl methacrylate		
1634044	Methyl tertiary butyl ether	75569	Propylene oxide
101144	4,4-Methylene bis (2-chloroaniline)	75558	1,2-Propylenimine (2-Methyl aziridine)
75092	Methylene chloride (Dichloromethane)	91225	Quinoline
		106514	Quinone
101688	Methylene diphenyl diisocyanate (MDI)	100425	Styrene
		96093	Styrene oxide
101779	4,4-Methylenedianiline	1746016	2,3,7,8-Tetrachlorodibenzo-p-dioxin
91203	Naphthalene		
98953	Nitrobenzene		

CAS Number	Chemical
79345	1,1,2,2-Tetrachloroethane
127184	Tetrachloroethylene (Perchloroethylene)
7550450	Titanium tetrachloride
108883	Toluene
95807	2,4-Toluene diamine
584849	2,4-Toluene diisocyanate
95534	o-Toluidine
8001352	Toxaphene (chlorinated camphene)
120821	1,2,4-Trichlorobenzene
79005	1,1,2-Trichloroethane
79016	Trichloroethylene
95954	2,4,5-Trichlorophenol
88062	2,4,6-Trichlorophenol
121448	Triethylamine
1582098	Trifluralin
540841	2,2,4-Trimethylpentane
108054	Vinyl acetate
593602	Vinyl bromide
75014	Vinyl chloride
75354	Vinylidene chloride (1,1-Dichloroethylene)
1330207	Xylenes (isomers and mixture)
108383	m-Xylenes
95476	o-Xylenes
106423	p-Xylenes
0	Antimony compounds
0	Arsenic compounds (inorganic including arsine)
0	Beryllium compounds
0	Cadmium compounds
0	Chromium compounds

CAS Number	Chemical
0	Cobalt compounds
0	Coke oven emissions
0	Cyanide Compounds[1]
0	Glycol ethers[2]
0	Lead compounds
0	Manganese compounds
0	Mercury compounds
0	Fine mineral fibers[3]
0	Nickel compounds
0	Polycyclic organic matter[4]
0	Radionuclides (including radon)[5]
0	Selenium compounds modifications

NOTE: For all listings above that contain the word "compounds" and for glycol ethers, the following applies: Unless otherwise specified, these listings are defined as including any unique chemical substance that contains the named chemical (i.e., antimony, arsenic, etc.) as part of that chemical's infrastructure.

1. X'CN where X = H' or any other group where a formal dissociation may occur, for example KCN or $Ca(CN)_2$

2. Includes mono- and di- ethers of ethylene glycol, diethylene glycol, and triethylene glycol $R-(OCH_2CH_2)n-OR'$ where
n = 1, 2, or 3
R = alkyl or aryl groups
R' = R, H, or groups which, when removed, yield glycol ethers with the structure: $R-(OCH_2CH)_n-OH$. Polymers are excluded from the glycol category. (See Modification.)

3. Includes mineral fiber emissions from facilities manufacturing or processing glass, rock, or slag fibers (or other mineral-derived fibers) of average diameter 1 micrometer or less.

4. Includes organic compounds with more than one benzene ring, and a boiling point greater than or equal to 100 °C.

5. A type of atom that spontaneously undergoes radioactive decay.

Modifications to the 112(b)(1) Hazardous Air Pollutants

Glycol Ethers

On January 12, 1999 (64FR1780), the EPA proposed to modify the definition of glycol ethers to exclude surfactant alcohol ethoxylates and their derivatives (SAED). On August 2, 2000 (65FR47342), the EPA published the final action. This action deletes each individual compound in a group called the surfactant alcohol ethoxylates and their derivatives (SAED) from the glycol ethers category in the list of hazardous air pollutants (HAP) established by section 112(b)(1) of the Clean Air Act (CAA). Under section 112(b)(3)(D) of the CAA, EPA may delete specific substances from certain listed categories, including glycol ethers. To implement this action, EPA is revising the definition of glycol ethers to exclude the deleted compounds. This action is also making conforming changes with respect to designation of hazardous substances under the Comprehensive Environmental Response, Compensation, and Liability Act (CERCLA). These final rules are being issued by EPA in response to an analysis of potential exposure and hazards of SAED that was prepared by the Soap and Detergent Association (SDA) and submitted to EPA. Based on this information, EPA has made a final determination that there are adequate data on the health and environmental effects of these substances to determine that emissions, ambient concentrations, bioaccumulation, or deposition of these substances may not reasonably be anticipated to cause adverse human health or environmental effects. All information associated with this rulemaking is available at EPA's Air and Radiation Docket and Information Docket, Room M1500, U.S. Environmental Protection Agency, 401 M Street, SW, Washington, DC 20460. The docket is an organized and complete file of all the information considered by the EPA in the development of this rulemaking. The docketing system is intended to allow members of the public and industries involved to readily identify and locate documents so that they can effectively participate in the rulemaking process. Along with the proposed and promulgated standards and their preambles, the contents of the docket will serve as the record in the case of judicial review. (See section 307(d)(7)(A) of the CAA.) An index for each docket, as well as individual items contained within the dockets, may be obtained by calling (202) 260-7548 or (202) 260-7549. Alternatively, docket indexes are available by facsimile, as described on the Office of Air and Radiation, Docket and Information Center web site at http://www.epa.gov/oar/docket. A reasonable fee may be charged for copying docket materials. A useful reference for the glycol ether category is linked below.

Toxics Release Inventory: List of Toxic Chemicals Within the Glycol Ethers Category (December 2000).

Caprolactam

On July 19, 1993, EPA received a petition from AlliedSignal, Inc., BASF Corporation, and DSM Chemicals North America, Inc. to delete caprolactam (CAS No. 105-60-2) from the hazardous air pollutant list in Section 112(b)(1), 42 U.S.C., Section 7412(b)(1). A Notice of Receipt was published (58FR45081, August 26, 1993) noting that the data filed were adequate to support decision making. After a comprehensive review of the data submitted, the EPA published a proposal to delist caprolactam (60FR48081, September 18, 1995). In order to help address public concern,on March 13, 1995, EPA executed two detailed agreements with AlliedSignal concerning the Irmo, South Carolina, manufacturing facility and another facility located in Chesterfield, Virginia, copies of which are included in the public docket for this rulemaking. AlliedSignal agreed that, if caprolactam was delisted pursuant to the proposal, AlliedSignal would install emissions controls, which EPA believed would be equivalent to the controls which would have been required had EPA issued a standard to control these sources under Section 112. The agreed emissions controls are incorporated in federally enforceable operating permits for the affected facilities, and will be in place years earlier than controls would have otherwise been required. In addition, AlliedSignal has agreed to establish a citizen advisory panel concerning the Irmo facility in order to improve communications with the community and to ensure that citizens have an ongoing role in implementation of the agreed emission reductions. The public requested a public hearing. On November 28, 1995, the EPA published a notice of public hearing and an extention of the comment period (60FR58589). After considering all public comments, the EPA published a final rule delisting caprolactam (61FR30816, June 18, 1996). All information associated with this rulemaking is located in Docket Number A-94-33 at the Central Docket Section (A-130), Environmental Protection Agency, 401 M St. SW., Washington, DC 20460; phone 202-260-7548, fax 202-260-4400, e-mail a-and-r-docket@epamail.epa. gov.

The docket includes a complete index to all papers filed in this docket, a copy of the original petition, comments submitted, and additional materials supporting the rule. A reasonable fee may be charged for copying. The docket may be inspected in person between 8:00 a.m. and 4:30 p.m. on weekdays at EPA's Central Docket Section, West Tower Lobby, Gallery 1, Waterside Mall, 401 M St., SW, Washington, DC 20460.

Hydrogen Sulfide

A clerical error led to the inadvertent addition of hydrogen sulfide to the Section 112(b) list of Hazardous Air Pollutants. However, a Joint Resolution to remove hydrogen sulfide from the Section 112(b)(1) list was passed by the Senate on August 1, 1991 (Congressional Record page S11799), and the House of Representatives on November 25, 1991 (Congressional Record pages H11217-H11219). The Joint Resolution was approved by the President on December 4, 1991. Hydrogen Sulfide is included in Section 112(r) and is subject to the accidental release provisions. A study (see citation below) was required under Section 112(n)(5).

Hydrogen Sulfide Air Emissions Associated with the Extraction of Oil and Natural Gas,
EPA-453/R-93-045,
NTIS (publication # PB94-131224, $36.50 hard copy, $17.50 microfiche).

National Technical Information Services (NTIS)
5285 Port Royal Road
Springfield, VA 22161
703-487-4650 800-426-4791
703-487-4807 8:30-5:30 EST M-F

Appendix

Air Pollution and Your Health

St. Louis Community Air Project (CAP) knows that your health and your family's health are important. It is our goal to reduce the risks to your health by identifying and reducing air pollutants. This will take the cooperation and hard work of everyone. We all contribute to air pollution and we can all do something about it. Visit the St. Louis CAP "What I Can Do" page at http://www.stlcap.org/whatsnew.htm for more information.

The St. Louis CAP Partnership spent several monthly meetings learning about health risk and how the U.S. Environmental Protection Agency calculates cancer and non-cancer health risks. This education process prepared the Partnership to address risk issues concerning:

- Duration of Exposure
- Adopting a Target Level of Risk
- Additivity of Cancer Risk
- Data Evaluation of the Monitors

The Risk Development Team, much like the Community Involvement Team, met outside of the regular monthly Partnership meeting to work on establishing the project's health benchmarks. This small group of Partners developed a list of pros and cons for the different options and offered recommendations to the full Partnership.

David Shanks made the presentation to the Partnership at the February 2001 meeting. and the partners discussed and voted on each recommendation. On every issue the Team and the full Partnership chose what came to be known as the "smorgasbord" approach—that is, to include as much information as possible when our monitoring data is analyzed. Based on these decisions, EPA and MDNR developed the official health benchmarks for the project.

These benchmarks will help the Partnership analyze and respond to the monitoring data. To learn more about the recommendations accepted by the Community Air Project Partnership as well as the methodology adopted for establishing health benchmarks, please review the following documents.

St. Louis Community Air Project Health Benchmark Recommendations

Risk Development Team Recommendations, presented to and accepted by the Community Air Project Partnership, Feb. 29, 2001.

The CAP Partnership will use the following recommendations to set health benchmarks for each of the chemicals we are monitoring. If a chemical we are monitoring poses a health risk according to our health benchmark, the Partnership will work with the community to reduce the risk.

Recommendation Re: Target Level of Cancer Risk

The Risk Development Team recommends a cancer risk level of 1 in 100,000 as the level of risk that would trigger additional activities to reduce exposure.*

Options for target level of cancer risk were:

1. 1 in 1,000,000

2. 1 in 100,000 or 10 in 1,000,000

3. 1 in 10,000 or 100 in 1,000,000

The pros of setting the target level at 1 in 100,000 are:

- It is consistent with most U.S. EPA regulatory programs.
- It allows us to actually estimate the cancer risk for more compounds. (Technology cannot detect most air pollutants at a concentration equal to a cancer risk of 1 in 1,000,000).

The cons of setting the target level at 1 in 100,000 are:

- It does not represent the most protective level for human health, which is 1 in 1,000,000.

*This is an EPA ranking system that classifies the likelihood that a chemical causes cancer in humans. The classification is based on both human and animal studies. Group A are human carcinogens, Group B are probable human carcinogens, and Group C are possible human carcinogens.

Appendix

Examples of Available Risk Reduction and Pollution Prevention Programs

The resources listed below are only examples of reduction and pollution prevention programs. A more comprehensive list of resources is now under development. Please check the Manual web site for updates to this resource list.

This list contains resources and programs for addressing a wide range of environmental issues, not just outdoor air issues. This more comprehensive list is provided as a resource for community partnerships that may want to address more than outdoor air issues.

Please note that some of the following links are not a part of the epa.gov domain. These links provide additional information that may be useful or interesting and are being provided consistent with the intended purpose of this EPA document. However, EPA cannot attest to the accuracy of information provided by these links or any other linked site. Providing links to a non-EPA web site does not constitute an endorsement by EPA or any of its employees of the sponsors of the site or the information or products presented on the site. Also, be aware that the privacy protection provided on the epa.gov domain may not be available at the external link.

PROGRAM TYPE	PROGRAM DESCRIPTION	WEB SITE OR POINT OF CONTACT
For Large Businesses and Public Facilities	**Identifying Pollution Prevention Opportunities** Encourage large chemical, refining, and manufacturing facilities to institute voluntary pollution prevention programs. Encourage companies to conduct audits to identify pollution prevention opportunities. Identify national industry sector leaders to use as benchmarks for local companies. Organize a community team with independent expertise to help facilities identify pollution prevention opportunities.	http://cfpub.epa.gov/clearinghouse/index.cfm http://www.epa.gov/compliance/assistance/sectors/ index.html http://www.epa.gov/opptintr/p2home/resources/ index.htm
For Small Businesses	**Design for the Environment Program** EPA partnership program working with individual industry sectors to compare and improve the performance and human health and environmental risks and costs of existing and alternative products, processes, and practices. DfE partnership projects promote integrating cleaner, cheaper, and smarter solutions into everyday business practices. Partnership programs include auto refinishing, printing and publishing, and dry-cleaning businesses.	http://www.epa.gov/dfe/projects/auto/index.htm http://www.epa.gov/dfe/projects/flexo/index.htm http://www.epa.gov/dfe/projects/gravure/index.htm http://www.epa.gov/dfe/projects/litho/index.htm http://www.epa.gov/dfe/projects/screen/index.htm http://www.epa.gov/dfe/projects/garment/index.htm

PROGRAM TYPE	PROGRAM DESCRIPTION	WEB SITE OR POINT OF CONTACT
For Small Businesses *(continued)*	**Environmental Results Program** An innovative program designed to assist businesses to improve their performance and address environmental problems. In the Environmental Results Program communities and regulating agencies can combine resources to educate businesses about their environmental impacts and obligations, help them to certify their compliance, and track them to evaluate their environmental performance.	http://www.epa.gov/permits/masserp.htm http://www.epa.gov/compliance/incentives/innovations/programresults.html
	Green Business Program Organize a program like the Bay Area Green Business Program, a partnership of community organizations, environmental agencies, professional associations, waste management agencies, and utilities to work together to recognize and assist businesses that operate in an environmentally friendly manner.	http://www.abag.ca.gov/bayarea/enviro/gbus/gb.html
	Businesses for the Bay Create a voluntary organization of businesses, like the Businesses for the Bay organization in the Chesapeake Bay watershed, committed to helping each other implement pollution prevention in daily operations and reduce releases of chemical contaminants and other wastes to your watershed.	http://www.chesapeakebay.net/b4bay.htm
	Stationary Source Pollution Prevention Fact Sheets for Communities and Small Businesses Multiple fact sheets on topics such as metal operations, electroplating, autobody paint shops, and printers. Includes information designed to help communities identify pollution prevention and reduction opportunities for small businesses. Designed to provide concrete assistance to help small shops implement easy pollution prevention measures and reduce releases of air toxics. Fact sheets now in final production.	Contact Amanda Aldridge at: aldridge.amanda@epa.gov
For Schools	**Tools for Schools** Voluntary, easy-to-use resource kit to help schools identify, remedy, and prevent indoor air quality problems in a cost-effective manner. Schools implement a range of specific guidelines emphasizing reduced pesticide exposure, use, and safe chemical storage, proper ventilation, and more.	http://www.epa.gov/iaq/schools/

PROGRAM TYPE	PROGRAM DESCRIPTION	WEB SITE OR POINT OF CONTACT
For Schools *(continued)*	**Clean School Bus USA** Brings together partners from business, education, transportation, and public health organizations to work to reduce pollution from public school buses. Includes policies and practices to eliminate unnecessary idling, retrofit buses with newer control technologies, and replace older buses.	http://www.epa.gov/otaq/schoolbus/
For Mobile Sources	**Voluntary Diesel Retrofit Program** Develop a program to retrofit older diesel engines with modern emission control technology. Enlist private and/or public fleets for participation.	http://www.epa.gov/otaq/retrofit/
	Anti-Idling Campaigns Develop education campaign and administrative policies to discourage vehicle idling in areas where people congregate.	
	Vehicle Engine and Maintenance Campaigns Sponsor a campaign to encourage proper vehicle and engine maintenance. Could involve a "tune your car today" at a local garage, checklists, and parts giveaways for do-it-yourselfers, etc.	
For Community Surface Waters	**Fish Consumption Surveys and Advisories** Perform surveys to determine whether there should be more fish/wildlife consumption advisories. Make advisories widely available to the public by print, radio, or television in multiple languages with an emphasis on subpopulations with high expected consumption.	http://www.epa.gov/waterscience/fish/
	Watershed Protection A Watershed Protection Approach is a strategy for effectively protecting and restoring aquatic ecosystems and protecting human health. This strategy has as its premise that many water quality and ecosystem problems are best solved at the watershed level rather than at the individual waterbody or discharger level. Major features of a Watershed Protection Approach are: targeting priority problems, promoting a high level of stakeholder involvement, designing integrated solutions that make use of the expertise and authority of multiple agencies, and measuring success through monitoring and other data gathering.	http://www.epa.gov/owow/watershed/index2.html

PROGRAM TYPE	PROGRAM DESCRIPTION	WEB SITE OR POINT OF CONTACT
For Community Surface Waters *(continued)*	**Mercury Reduction in Hospitals** Help hospitals comply with new requirements by providing information and assistance. Encourage hospitals to eliminate mercury sources such as thermometers. Conduct education programs for citizens and hospital staff about mercury reduction.	http://www.noharm.org/mercury/issue
	Household Mercury Thermometer Exchanges Sponsor a trade-in program that provides citizens with new, non-toxic thermometers in exchange for mercury thermometers to reduce risk of mercury contamination in homes and to reduce the risk of water contamination and outdoor air pollution due to improper disposal.	http://www.noharm.org/mercury/issue
	National Estuary Program The National Estuary Program is designed to encourage local communities to take responsibility for managing their own estuaries. Each NEP is made up of representatives from federal, state, and local government agencies responsible for managing the estuary's resources, as well as members of the community — citizens, business leaders, educators, and researchers. These stakeholders work together to identify problems in the estuary, develop specific actions to address those problems, and create and implement a formal management plan to restore and protect the estuary.	http://www.epa.gov/owow/estuaries/about2.htm
	Coastal America Coastal America is a unique partnership of federal agencies, state and local governments, and private organizations. The partners work together to protect, preserve, and restore our nation's coasts.	http://www.coastalamerica.gov/
For Community Homes	**Develop a Community Campaign using Home*A*Syst** Home*A*Syst is an environmental risk assessment guide for the home that helps homeowners identify risks and take actions to protect health and the environment. Organize a community education campaign using the Home*A*Syst program and materials.	http://www.hud.gov/offices/lead/helpyourself/index.cfm
	Radon "Test and Repair" Campaigns Enlist citizens to test their home for radon and provide information and assistance to correct the problem if radon levels are unacceptably high.	http://www.epa.gov/iaq/radon/

PROGRAM TYPE	PROGRAM DESCRIPTION	WEB SITE OR POINT OF CONTACT
For Community Homes *(continued)*	**Home Consumer Products Education Campaigns** Educate citizens in practices they can adopt such as proper solvent storage, vehicle operation tips, landscaping and yard-care options to minimize use of pesticides and polluting equipment, use of lower toxicity home products, etc.	http://www.epa.gov/oar/oaqps/peg_caa/ pegcaa07.html http://www.epa.gov/reg3esd1/garden/index.htm http://www.epa.gov/greenkit/landscape.htm
	Low Emission Gas Can Exchanges Emissions from portable fuel cans present a significant source of exposure to gaseous toxics such as benzene, especially if the can is stored inside a dwelling or attached garage. Encourage citizens to exchange their old-style containers for new ones meeting higher standards.	http://www.arb.ca.gov/msprog/spillcon/gascanfs/ gascanfs.htm
	Campaign for a Lead Safe America Protect community children with an education and testing program to reduce exposure to lead in homes and soil.	http://www.hud.gov/offices/lead/outreach/ communityoutreach.cfm#leadsafehome
	Lead in Drinking Water Campaigns Approximately 20% of human exposure to lead is attributable to lead in drinking water. Provide education about ways to reduce exposure to lead in drinking water.	http://www.epa.gov/OGWDW/Pubs/lead1.html
	National Asthma Public Education and Prevention Campaigns Conduct an education campaign in schools and homes to reduce asthma and to increase the asthma awareness and asthma triggers.	http://www.epa.gov/asthma/
	Integrated Pest Management Programs Integrated pest management (IPM) uses habitat modification, biological controls, and chemical controls. IPM protects people from noxious pests and toxic pesticides. Conduct a community Integrated Pest Management (IPM) Education Campaign.	http://schoolipm.ifas.ufl.edu/ http://www.epa.gov/pesticides/
	Household Hazardous Waste Collections Exposure to hazardous household materials can be significantly reduced by collecting old and unused products and disposing of them properly. Conduct a neighborhood drive to collect pesticides, coolants, lubricants, solvents, and other hazardous products, some of which are now banned due to their toxicity.	http://www.epa.gov/epaoswer/non hw/muncpl/hhw.htm

PROGRAM TYPE	PROGRAM DESCRIPTION	WEB SITE OR POINT OF CONTACT
For Community Homes *(continued)*	**Smoke-Free Homes and Cars Campaigns** Making homes and cars smoke-free are an easy and proven ways to protect nonsmokers from secondhand smoke exposure. Conduct a smoke-free campaign using existing materials, including television, radio, and print public service announcements (PSAs), smoke-free home brochures, and the toll-free pledge number, and other materials.	http://www.epa.gov/smokefree/index.html

Appendix

Toxicity Values and Toxicity Sources: Background for the Collection of Information to Calculate Screening-Level Concentrations

Examples of toxicity values for calculating screening-level concentrations

To complete step 2 of Figure 10-2, the technical team collects information that describes each community chemical's potential to cause a long-term, adverse health effect. Depending on what it learns about what kinds of long-term adverse health effects each community chemical may cause, the technical team may have to find one or, for some chemicals, two types of toxicity values. If a chemical has a potential to cause cancer, the technical team needs to find a cancer toxicity value for the chemical. If a chemical has a potential to cause a non-cancer, long-term health effect, the technical team needs to find a non-cancer toxicity value for the chemical. If a chemical has a potential to cause cancer as well as some other non-cancer, long-term health effect, the technical team needs to find both types of toxicity values for the chemical.

Examples of toxicity values for cancer-causing chemicals include the cancer potency slope factor (CSF or simply SF) and the cancer unit risk (UR) estimate.

A **cancer potency slope factor** provides an estimate of increased cancer risk from a lifetime exposure to a given chemical. This estimate is usually derived from a study of oral exposure to the chemical of interest. The unit of expression for the oral cancer potency slope factor is a portion of a population affected per milligram (one-thousandth of a gram) of chemical per kilogram of body weight per day.

A **cancer unit risk estimate** also provides an estimate of increased cancer risk from a lifetime exposure to a given chemical. This estimate is usually derived from an inhalation exposure study. The unit of expression for the

cancer unit risk estimate for inhalation exposure is a portion of a population affected per microgram (one millionth of a gram) of chemical per unit (cubic meter) of air breathed. It describes an increased cancer risk that may result from continuous inhalation exposure to a chemical at this air exposure concentration.

The cancer unit risk estimate is the preferred cancer toxicity value for use in the cancer screening-level concentration equation in step 3 of Figure 10-2.

Examples of toxicity values for non-cancer-causing chemicals include the inhalation reference dose, the reference concentration, the minimal risk level, and the reference exposure level.

A **reference concentration (RfC)** is an estimate (with uncertainty spanning perhaps an order of magnitude) of continuous level of inhalation exposure of the human population (including sensitive subgroups) that is likely to be without an appreciable risk of adverse effects during a lifetime. EPA generally applies uncertainty and modifying factors to NOAELs, LOAELs, or benchmark concentrations to set RfCs. EPA uses these factors to account for specified limitations of the available chemical toxicity information. The inhalation reference concentration is the preferred non-cancer toxicity value for use in the non-cancer screening-level concentration equation in step 3 of Figure 10-2. The unit of expression for the RfC is milligrams of chemical per cubic meter of air breathed. The unit of expression for the RfC can be converted to micrograms of chemical per cubic meter of air breathed by multiplying the former by 1,000 μg/mg.

A **minimal risk level (MRL)** is an estimate of the daily human exposure level to a hazardous (toxic) substance that has a low risk of adverse non-cancer health effects over a specified duration of exposure. The federal government's Agency for Toxic Substance and Disease Registry (ATSDR) calculates MRLs using an approach similar to the one used by EPA to develop its estimates of reference concentrations and reference doses. The unit of expression for an oral MRL is milligrams of chemical per kilogram of body weight per day. The unit of expression for an inhalation MRL is parts per million (ppm) of a chemical in air. As noted above, the RfC is the preferred non-cancer toxicity value for use in the non-cancer SLC equation in step 3 of Figure 10-2. To translate from ppm to units of mg/m³ (per RfC convention), the MRL is multiplied by the chemical's MW (mg/mMole) divided by 24.45 (mMole per m³ at 25 °C and 1 atmosphere pressure).

A long-term (chronic) **reference exposure level (REL)** is an airborne chemical concentration that would pose no significant health risk to individuals indefinitely exposed to that level. The California Environmental Protection Agency (CAL/EPA) calculates the REL, using an approach similar to the one used by EPA to develop estimates of its reference concentrations. The unit of expression for a REL is micrograms of chemical per cubic meter of air breathed. As noted above, the RfC is the preferred non-cancer toxicity value for use in the non-cancer SLC equation in step 3 of Figure 10-2.

Deriving inhalation values from oral values

When an inhalation toxicity value is not available for a chemical of interest, but an oral exposure route value is available, the Technical Team may want to consider extrapolation from the oral to an inhalation value. The Agency considers data from other routes potentially useful to derivation of an inhalation value only when respiratory tract effects and/or "first-pass" effects (a pharmacologic phenomenon) can be ruled out. (See section 4.1.2, USEPA, 1994. Methods for Deriving Inhalation Reference Concentrations and Application of Inhalation Dosimetry.) First-pass effects refer to the metabolism that can take place in the portal-of-entry tissue (e.g., the respiratory tract), prior to entry into the systemic circulation. A first-pass effect can alter the disposition of the parent and metabolite chemicals, thereby affecting the magnitude of the dose to remote target tissues in a route-dependent fashion. In the absence of data to determine dosimetry via inhalation, when a chemical is thought to be susceptible to first-pass effects (e.g., metabolized), or where a potential for portal-of-entry effects is indicated but not well characterized (e.g., respiratory toxicity after acute

exposures), then route-to-route extrapolation for derivation of an RfC or inhalation unit risk (IUR) is not appropriate.

For chemicals for which respiratory toxicity and first-pass effects can be ruled out, route-to-route becomes a possibility. Methods for this extrapolation range in accuracy and therefore, inherent uncertainty, with the simplest approach using default absorption values for each exposure route appropriate to the chemical class in question. The general equations for the simplest approach are as follows.

Equation 1. Extrapolation from an oral cancer SF to an inhalation unit risk (IUR) estimate

IUR (m³/μg) = CSF (kg-day/mg) * 20 m³ /day * 1/70 kg * 1mg/10³ μg

where:

IUR is an inhalation cancer unit risk estimate for the chemical

CSF is the oral cancer potency slope factor for a chemical

20 m³ is an assumption of the daily inhalation intake

70 kg is an assumption of the body weight

Equation 2. Extrapolation from an oral RfD to an inhalation noncancer reference value

Inhalation RfV (mg/m³) = RfD (mg/kg-day) * 1/day/ 20 m³ * 70 kg

where:

Inhalation RfV is an inhalation version of the oral non-cancer reference dose estimate for the chemical

RfD is the oral reference dose for a chemical

20 m³ is an assumption of the daily inhalation intake

70 kg is an assumption of the body weight

It is noted that the paired values 700kg and 20 m³/day are standard defaults intended to be protective of humans. The mean daily inhalation intake, presuming a variety of activity levels (rest to heavy activity) during a 24-hour period, would be less than 20 m³ for all age groups. Given the size of other sources of uncertainty in the route-to-route extrapolation, the small difference (e.g., 15%) that might be obtained through replacement of this default pair of values in the equations, with age-specific body weight and inhalation intake values, is considered insignificant.

References for the discussion above can be found on the web site of EPA's National Center for Environmental Assessment at http://cfpub.epa.gov/ncea/cfm/nceapubtopics.cfm?ActType=PublicationTopics.

Sources for chemical toxicity information

The following are sources of chemical toxicity information and chemical toxicity values. The accompanying text and source comparison Table H-1 describe each of these sources relative to the types and quality of information provided. Please see the discussion of the choice of toxicity information sources in the Overview section of this Manual.

- EPA's Office of Air Quality Planning and Standards (OAQPS) Air Toxics Dose-Response Database: http://www.epa.gov/ttn/atw/toxsource/summary.html

- EPA's Integrated Risk Information System (IRIS): http://www.epa.gov/ngispgm3/iria/subst/index.html

- EPA's Health Effects Assessment Summary Tables (HEAST) hard copy prepared by EPA's National Center for Environmental Assessment (NCEA) for EPA's Office of Solid Waste and Emergency Response (OSWER) for use at contaminated work sites.

- Agency for Toxic Substances and Disease Registry's (ATSDR) Minimal Risk Levels (MRLs): http://www.atsdr.cdc.gov/mrls.html

- California's Office of Environmental Health Hazard Assessment (OEHHA) or CAL/EPA Toxicity Criteria Database: http://www.oehha.ca.gov/risk/chemicalIDB/index.asp

- EPA's High Production Volume Toxicity Database : web site available in the future

OAQPS's Air Toxics Dose-Response Database is a compilation of toxicity values obtained from multiple sources both within and outside EPA. EPA uses these toxicity values in the National Air Toxics Assessment (NATA), which range from national- to local-scale applications. Toxicity values in OAQPS Air Toxics Dose-Response Database include those from IRIS and other EPA toxicity information databases, as well as those from ATSDR and CAL/EPA. Toxicity values describe toxic effects for chemicals for inhalation as well as for other exposure pathways. Toxicity values in the Air Toxics Dose-response Database include cancer unit risk estimates, cancer potency slope factors, RfCs, MRLs, and RELs.

The Air Toxics Dose-Response Database contains toxicity information on approximately 190 chemicals and chemical classes. Some chemicals/classes of chemicals in the Air Toxics Dose-Response Database may have toxicity information from more than one source. These toxicity information sources vary in the methodology used to develop their toxicity values, in their selection of sources of information used to develop toxicity values, in their characterizations of uncertainty, and in their levels of peer review. The Air Toxics Dose-Response Database provides a default order of preference (a priority or hierarchy) for different types of toxicity values when more than one type of information is available. OAQPS notes that changes to the hierarchy may be appropriate on a chemical-by-chemical basis.

EPA's **Integrated Risk Information System (IRIS)** is a high-quality, peer reviewed, and frequently updated chemical toxicity information source. Its ease of accessibility, chemical search capacity, and more than 500 chemicals currently within its database make it a good source of toxicity information and toxicity values. IRIS is EPA's official repository of Agency-wide consensus chronic human health toxicity information and toxicity values.

IRIS presents each of its chronic human health toxicity values for the inhalation exposure pathway as reference concentrations (RfCs) for non-cancer chemicals and cancer inhalation unit risk (UR) estimates for cancer chemicals. EPA updates IRIS monthly to provide consistent, up-to-date chemical toxicity information. The Partnership can access information through IRIS's home page. It can access information on IRIS chemicals by highlighting and selecting each chemical's name or CAS number.

EPA's **Health Effects Assessment Summary Tables (HEAST)** provide preliminary EPA chemical toxicity information and toxicity values. The database, prepared by EPA's National Center for Environmental Assessment (NCEA), consolidates toxicity information for chemicals of interest to Superfund, the Resource Conservation and Recovery Act, and EPA in general. The degree of peer review the toxicity information undergoes prior to its entry into HEAST is less than that for information in IRIS. The HEAST database currently contains toxicity information for over 500 chemicals. Unless otherwise stated, EPA considers the toxicity values for the chemicals in HEAST to be provisional. Although the information in HEAST has concurrence of individual

Table H-1.
A Comparison of Toxicity Information Sources

SOURCE	NUMBER OF CHEMICALS	TYPES OF DATA	UNITS	UPDATE FREQUENCY	LEVEL OF PEER REVIEW	COMMENTS
EPA IRIS	over 500	RfD RfC SF UR	mg/kg/day mg/m^3 mg/kg/day μg/L (water) μg/m^3 (air)	Monthly	Values have received Agency-wide consensus	RfD value is for chronic oral exposure only. RfC value is for chronic inhalation exposure only. Slope factor is for oral exposure only. Unit risk is available for drinking water and inhalation.
EPA HEAST	483 RfC (non-cancer) 181 (cancer)	RfD RfC SF UR	mg/kg/day mg/m^3 update (mg/kg/day)$^{-1}$ μg/L (oral) μg/m^3 (inhalation)	Last update was July 1997	Have undergone some level of EPA review but have not received final EPA approval	Consisting almost entirely of provisional risk HEAST assessment information. Subchronic and chronic RfD values are provided. Subchronic and chronic RfC values are provided. Oral and inhalation slope factors are provided. Oral and inhalation unit risk values are provided.
CAL/EPA - Toxicity Criteria Database	388	SF UR REL RfC	(mg/kg-day)$^{-1}$ (μg/m^3)$^{-1}$ μg/m^3 μg/m^3	Period-ically	RELs undergo review by the public and the Scientific Review Panel (SRP). Cancer potency values were select-ed from a heirarchy list of sources.	Oral and inhalation slope factors are provided. Inhalation unit risk is provided. Reported RELs are for acute and chronic exposure. RELs can be used as surrogates for RfC values when RFCs are not available. RFC values are provided in chemical summaries.
ATSDR	138	MRL	mg/m^3 (particles) ppm (gases)	Period-ically	Rigorous review process Internal and external peer review	MRLs were not based on cancer effects. MRLs were derived using no-observed-adverse-effect-level / uncertainty factor approach Chronic, acute, and intermediate MRLs are provided. MRLs can be used as surrogate for RfC when RfC values are not available.
OAQPS Air Toxics Dose-Dose-Response Database	188	SF UR REL RfC	(mg/kg-day)$^{-1}$ (μg/m^3)$^{-1}$ μg/m^3 μg/m^3	Period-ically	Pulled toxicity values from multiple sources and ranked according to data heirarchy	Updated on a periodic basis every 3-6 months. Values are presented in accordance with a default hierarchy of preferred sources, with preference given to those reflective of current knowledge, sound scientific basis, and external peer review. OAQPS notes that deviation from the default hierarchy may be appropriate on a chemical-specific basis. As available and consistent with the default tiered hierarchy, an EPA or CAL/EPA inhalation unit risk value, a chronic RfC, REL, or inhalation MRL, and a chronic RfD or oral MRL are provided.

EPA Program Offices and is supported by Agency references, there is not sufficient peer review of the information for it to be recognized as Agency-wide consensus information. Like IRIS, the HEAST database provides chronic non-cancer toxicity values as reference concentrations (RfCs) and reference dose (RfDs). The HEAST provides cancer toxicity values as potency slope factors and unit risk estimates for the inhalation exposure pathway. HEAST information is currently available in hard-copy format only. The date of the most current version of HEAST is July 1997.

The Agency for Toxic Substances and Disease Registry (ATSDR) within the United States Department of Health and Human Services provides toxicity information and non-cancer toxicity values (known as minimal risk levels or MRLs) for selected chemicals. ATSDR provides its information for chemicals found on the EPA's Superfund National Priorities List. ATSDR does not currently provide cancer toxicity values for cancer-causing chemicals. Currently ATSDR provides non-cancer toxicity information and toxicity values (MRLs) for approximately 150 chemicals or chemical groups. The MRL is an estimate of a daily human exposure to a chemical that has a low potential to cause adverse, non-cancer effects over a specified duration of exposure. ATSDR sets its MRLs below levels that, based on current information, might cause adverse, non-cancer human health effects in the most sensitive of exposed groups. The ATSDR sets its MRLs for various exposure durations and exposure pathways. The MRL of most interest to the Partnership's community air screening-level concentration setting process is the MRL for chronic (365 day or longer) exposure durations for the inhalation exposure pathway.

The method ATSDR uses to develop its MRLs is similar to the one used by EPA to develop its reference dose (RfD) and reference concentration (RfC) toxicity values. Proposed MRLs undergo a comprehensive peer review process. Four separate groups take part in the MRL review process. These groups include the Health Effects/ MRL Work Group within ATSDR's Division of Toxicology; an expert panel of external peer reviewers; an ATSDR-wide MRL Work Group with participation from other federal agencies including USEPA; and the public through the toxicological profile public comment period. ATSDR updates its MRLs periodically.

The **CAL/EPA's** (California Environmental Protection Agency's) Office of Environmental Health Hazard Assessment (OEHHA) maintains a toxicity information database. The database contains toxicity values on both cancer-causing and non-cancer-causing chemicals. Toxicity values include those that describe cancer potency slope factors, unit risk values, and chronic inhalation reference exposure levels (RELs). A chronic REL is an airborne chemical concentration that would pose no significant non-cancer health risk to individuals indefinitely exposed to that level. CAL/EPA bases its RELs solely on human health considerations from the best available human and animal toxicity information in the scientific literature. The California Air Pollution Control Officers' Association with consultation from CAL/OEHHA also develops chronic RELs, based on toxicity values previously established in readily available toxicity information sources. CAL/EPA's RELs and other toxicity values support decisions made in CAL/EPA's Air Toxics' "Hot Spots" program.

The CAL/EPA database currently contains toxicity information and toxicity values on approximately 400 chemicals. CAL/EPA updates its database periodically. As it should do with each of the other toxicity information sources, the Partnership should note the dates of latest update that the CAL/EPA database lists for toxicity values for community chemicals. There may be more current toxicity information available. The CAL/EPA web site provides a search engine that allows its users to search for toxicity information by chemical name. The web site provides an option for downloading specific cancer potency support documentation to view tables containing the entire database of cancer potency slope factor and cancer unit risk values. The database also provides an option of downloading related REL documentation to view the tables containing the entire database of RELs as well.

Steps for Calculating a Screening-Level Concentration

What steps can the technical team follow to calculate a screening-level concentration (SLC) value?

Figure 10-2 shows a three-step procedure the SLC Technical Team can use to calculate an air exposure SLC. The procedure for completing each of these three steps is described below. Two examples of SLC calculations are provided: one for a cancer risk-based SLC (for arsenic compounds), the other for a non-cancer risk-based SLC (for acrylic acid). Two alternate procedures are described in which the technical team might be able to derive an air exposure SLC when a chemical toxicity value is not available for one or more of its community chemicals.

Step 1: Set Community Risk Screening-Level Assumption Values

Set a cancer risk screening level and a non-cancer risk screening level for chemicals as appropriate, given their potential toxicities.
Example values: Additional lifetime risk of 1E-06 for a cancer risk screening level; a hazard quotient of 1 for a non-cancer risk screening level. See discussion describing the Partnership's choice of risk screening levels in the Initial Screen chapter of the Overview section.

Step 2: Collect Toxicity Information/Values

The Partnership should establish a toxicity information source hierarchy. The information source hierarchy is important to the consistency and transparency of the Partnership's toxicity information collection process.

Collect toxicity information.
There are two types of chemical toxicity values of interest: one for cancer, the other for non-cancer effects. The SLC Technical Team uses its cancer toxicity values (cancer potency slope factors/inhalation cancer unit risk estimates) to calculate air exposure SLCs for cancer-

causing chemicals. The technical team uses its non-cancer toxicity values to calculate air exposure SLCs for non-cancer-causing chemicals. Some chemicals found in the community will require calculations of both cancer and non-cancer SLCs. For these chemicals the technical team collects toxicity values for both cancer and non-cancer effects.

The technical team uses its toxicity information hierarchy to identify an order and uses toxicity data sources and values for its SLC calculations. An example of such an order of preference for non-cancer toxicity values would be: (1)RfC (an inhalation reference concentration), (2) REL (an inhalation reference exposure level), and (3) MRL (minimal risk level). Whichever values the Partnership chooses to use, it is important that the values are current.

Step 3: Calculate an Air Exposure Screening-Level Concentration

Use values for the cancer risk screening-level assumption and the cancer toxicity value assumption to calculate a chemical-specific cancer screening-level concentration.
Not all chemicals cause cancer. For those that may cause cancer, the SLC Technical Team calculates cancer screening-level concentrations. It uses the following equation and assumption values:

$$\text{SLC (cancer) } \mu g/m^3 = \frac{\text{RSL}}{\text{UR}}$$

where:

RSL = cancer risk screening level (e.g., 1.0E-06)

UR = chemical-specific inhalation unit risk estimate (per $\mu g/m^3$)

Use values for the non-cancer risk screening-level assumption and the non-cancer toxicity value assumption to calculate a chemical-specific non-cancer screening-level concentration.

Not all chemicals cause non-cancer, long-term health effects. For those that may cause non-cancer effects, the technical team calculates non-cancer screening-level concentrations (it refers to them as its non-cancer SLCs). It uses the following equations and assumption values:

SLC (non-cancer) $\mu g/m^3$ = SHQ * RfC * 1000 $\mu g/mg$

where:

SHQ = screening hazard quotient (e.g., 1.0)

RfC = Chemical-specific inhalation reference concentration (mg/m^3)

Compare a chemical's cancer SLC value to its non-cancer SLC value and use the most conservative (lowest) value as the chemical's air exposure screening-level concentration.

Not all chemicals will have toxicity values for both cancer and non-cancer effects. For those chemicals that have both cancer and non-cancer toxicity values, the technical team calculates cancer and non-cancer SLCs. The smaller of the two SLCs represents the more conservative, more protective screening-level concentration. This is the air exposure SLC that the technical team will use to compare to its estimates of ambient air concentrations for community chemicals.

What are some examples of SLC calculations?

The following two examples show how the technical team calculates its air exposure SLCs for cancer (the arsenic compounds example) and non-cancer (the acrylic acid example) chemicals. Toxicity values were found in OAQPS's Air Toxics Dose-Response Database (http://www.epa.gov/ttn/toxsource summary120202.html).

EXAMPLE 1: Arsenic Compounds (cancer SLC example)

SLC (cancer) $\mu g/m^3 = \dfrac{1.0\text{E-}06}{4.3\text{E-}03 \text{ per } \mu g/m^3(\text{ from IRIS})} = 2.3\text{E-}04 \ \mu g/m^3$

EXAMPLE 2: Acrylic acid (non-cancer SLC example)

SLC (non-cancer) $\mu g/m^3$ = 1 * 1.0E-03 mg/m^3
(from IRIS) * 1,000 $\mu g/mg$
= 1.0 $\mu g/m^3$

How can the SLC Technical Team calculate an SLC when toxicity information is not available?

Before considering the use of the following two alternate methods to derive SLCs, the technical team should exhaust **all** of its sources of toxicity information. Existing toxicity information for a chemical may not be in one source but may be available in another. However, if the technical team does not find the necessary toxicity information for one or more of its chemicals, it can use toxicity information for a surrogate chemical or a default chemical to derive the necessary SLCs.

To the extent possible, a chemical used as a surrogate should have a similar chemical structure and similar physical and chemical properties as the chemical of interest. For example, the Virginia Department of Environmental Quality recommends pyrene as a surrogate chemical for phenanthrene and benzo(g,h,i)pyrene. If either of these latter two chemicals were a community chemical for which the Partnership could find no toxicity information, it could use toxicity values for pyrene to calculate an air exposure SLC for either chemical. The technical team should make case-by-case determinations and get Partnership approval for chemicals to be used as surrogates for toxicity information.

If the technical team does not find the necessary toxicity information for one or more of its chemicals or chooses not to use information on a surrogate chemical, it can use toxicity information on a chemical it chooses (and gets Partnership approval) to be its default chemical. The technical team (Partnership) has at least four options to make its choice: (1) It can select its default chemical from among the most toxic of its community chemicals. (2) It can select its default chemical from among the least toxic of its community chemicals. (3) It can select its default chemical from among the moderately toxic of its community chemicals. Or (4) it can select as its default chemical a chemical that may not even be among those identified as community chemicals. The Partnership, with the advice of its SLC Technical Team, will make the final choice. The Partnership's choice will be based on how conservative it wants its air exposure screening-level concentrations for those community chemicals with little or no toxicity information.

Methods for Apportioning County Data

Methods for Apportioning County Data

Land use/land cover (LULC)—determine area of different land use categories in census tract and study area. Develop ratio of area of land use category in census tract to area of land use category in study area. Different LULC ratios can be used with different sources (e.g., using commercial LULC ratio with autobody refinishing). LULC data for standard 1:250,000 scale topographic maps can be obtained on the Internet at http://edc.usgs.gov/geodata under the LULC tab. Users should consult the online user's guide and the "read me" file for more information on these files. A user's guide that explains the structure of the LULC data files can be obtained at ftp://map.usgs.gov/pub/ti/LULC/lulcguide.

Users with access to GIS software should download files under the "land_use" heading. These files are in the geographic information retrieval and analysis system (GIRAS) format. This format uses arcs and polygons to define regions on the map.

Otherwise, users should download files in the composite theme grid (CTG) format under the "grid_cell" heading. The CTG format uses grid cells to define locations on a map. The grid cells are actually a regular point sample of the quad where the center point of each cell is 200 meters from other center points in adjacent cells. The cells are mapped to the Universal Transverse Mercator (UTM) projection, oriented in north-south, east-west directions, and sequenced by row from north to south, within each row, by column east to west. More information on these files can be found at the end of this write-up.

CTG files can be opened in any word-processing software. The best view seems to occur when the font has been changed to Courier 10 point, with left and right margins set to 0.9 inches. These files can be very large (thousands of pages) and can be complicated to analyze and manipulate.

Area—ratio of surface area of census tract to surface area of study area. Area data can be obtained using GIS or topographic maps.

Population—ratio of population of census tract to population of study area. Population data can be obtained from the U.S. Census Bureau.

Population density—ratio of population to area for census tracts. Sum population densities for entire study area. Determine ratio of population density for census tract to population density for study area. Population density data can be obtained from GIS, topographic maps, and the U.S. Census Bureau.

Roadway miles—ratio of total roadway miles in census tract to total roadway miles in study area. For use with mobile sources. Roadway mileage data can be obtained from the Department of Transportation, GIS, or topographic maps.

Composite Theme Grid (CTG) Data File Format

Digital data from all the overlays of a given quadrangle also are combined in a raster or grid cell format as a Composite Theme Grid (CTG) file.

CTG files are sequential and consist of fixed-length logical records, and with the exception of header records, all records are of identical internal format, one grid cell per logical record. The grid cells are actually a regular point sample. The attribute codes at the center point of each cell are recorded from each overlay. The points are oriented to the UTM projection and are usually spaced 200 m apart in both east-west and north-south directions. The cell records are first ordered in the file by row from north to south, then within each row, by column west to east.

Character Composite Theme Grid (CTG) File Format

A character-formatted (usually ASCII) CTG file consists of fixed-length 80-character (card image) logical records. There are two parts to the CTG file, first a header then cell records. All records, except the last header record with one text field, consist of fixed-length integer fields; each integer is coded as digits with leading blanks (i.e., right justified). The first five logical records of the file comprise the CTG map header. The header is followed by cell records, one grid cell per 80-character logical record.

In some cases a CTG file may be released without the map header contained in the file. In this case all records in the file are individual grid cell records, and the header information is supplied as a printed listing.

Character CTG Map Header

Record 1:

Bytes 1-10 = Number of rows;

11-20 = Total number of cells x 2;

21-30 = Number of columns;

31-35 = Meaningless field;

36-40 = Cell size (width and length) in meters;

41-45 = Number of overlays merged;

46-50 = Map type code (see below);

51-55 = Projection zone number;

56-60 = Map projection code (should be "1" for UTM);

61-70 = Scale of a plot at one mil per cell width; and

71-80 = Source date of the land use overlay.

Record 2:

Bytes 1-5   = Minimum column index;

6-10 = Minimum row index;

11-15 = Maximum column index;

16-20 = Maximum row index;

21-25 = Column index for SW control point;

26-30 = Row index for SW control point;

31-35 = Column index for NW control point;

36-40 = Row index for NW control point;

41-45 = Column index for NC control point;

46-50 = Row index for NC control point;

51-55 = Column index for NE control point;

56-60 = Row index for NE control point;

61-65 = Column index for SE control point;

66-70 = Row index for SE control point;

71-75 = Column index for SC control point; and

76-80 = Row index for SC control point.

Record 3:

Bytes 1-10 = Latitude of SW control point;

11-20 = Longitude of SW control point;

21-30 = Latitude of NW control point;

31-40 = Longitude of NW control point;

41-50 = Latitude of NC control point;

51-60 = Longitude of NC control point;

61-70 = Latitude of NE control point; and

71-80 = Longitude of NE control point.

Record 4:

Bytes 1-10 = Latitude of SE control point;

11-20 = Longitude of SE control point;

21-30 = Latitude of SC control point;

31-40 = Longitude of SC control point;

41-50 = UTM Easting value of west edge of cells;

51-60 = UTM Northing value of north edge of cells;

61-70 = File creation date (a Julian date); and

71-80 = Meaningless field.

Record 5:

Bytes 1-64   = Title (text characters); and

65-80 = Blank.

Some further explanation is needed for some of the elements in the CTG map header:

1. The map type code (in bytes 46-50 of the first record) indicates which overlays have been included in the CTG data file. The code is formed by the addition (in base 10) of the separate GIRAS map type codes for each of the overlays:

10   = Land Use and Land Cover;

02   = Political units;

04 = Census county subdivisions and SMSA tracts;

10 = Hydrologic units;

20 = Federal land ownership; and

40 = State land ownership.

For example, the map type code for a combination of the first four overlays above would be 17; all six overlays combined have a map type code of 77.

2. The UTM Easting and Northing values given in the fourth record (bytes 41-60) are in whole meters and are values for the west and north edges of the cells, rather than the center point of the first (northwest corner) cell. The Easting and Northing values for a given cell may be calculated thus:

Easting = (XORG CW/2) + (column index)*CW
Northing = (YORG+CW/2) (row index)*CW

where XORG and YORG are the Easting and Northing values in bytes 41-60 of the fourth header record, and CW is the cell width in bytes 36-40 of the first header record.

3. The control points usually define the 1_ x 2_ (for 1:250,000-scale base maps) or 30' x 1_ (for 1:100,000 scale base maps) quadrangle on which the overlay data are based. The latitude and longitude values are given as positive integers of the form DDDMMSS, where DDD is degrees, MM is minutes, and SS is seconds. West longitude values are given as positive numbers, increasing in value from east to west. The row and column values given for the control points are the indices for the cell whose center point is closest to the true position of the control point.

CTG Grid Cell Records

Each grid cell logical record of a standard character-formatted CTG data file is 80 characters in length and consists of nine decimal integers, right justified (with leading blanks) within fixed-length fields:

Bytes 1-3 = UTM zone number (this value should be the same in every record of a given CTG file); the first byte will always be a blank for zones in the northern hemisphere;

4-11 = UTM Easting value, in whole meters, for the sample point of the cell;

12-19 = UTM Northing value, in whole meters, for the sample point of the cell;

20 = Blank;

21-30 = Land Use and Land Cover attribute code;

31-40 = Political unit (FIPS State/county) code;

41-50 = USGS hydrologic unit code;

51-60 = Census county subdivision or SMSA tract code;

61-70 = Federal land ownership agency code; and

71-80 = State land ownership code.

If a given overlay category has not been included within the file, the codes for that category will be zero (0). Since some misregistration of map overlays occurs, some of the cells along the edges of the 1:250,000- or 1:100,000 scale quadrangle may have codes for some overlays, but not others (the "other" codes will be zero). The standard character CTG data file will have only those cell records for which at least one of the categories is coded. This means that, since the 1:250,000 and 1:100,000 scale quadrangles do not form perfect rectangles in the UTM projection (lines of latitude curve and lines of longitude converge), a variable number of cell records will exist for any given row or column.

Binary CTG Data File Format

Each logical record of a binary CTG file is either 32 or 52 bytes in length. A record consists of eight 32-bit (4 byte) binary integers in the following order:

Bytes 1-4 = Row index, where 1 is the index of the northernmost row and index numbers increase by one for each row moving south (NOTE, due to a processing error, CTG files in which the State ownership is not coded will have all zero row index numbers; the row index is then a function of the sequential position of the record within the file);

5-8 = Column index, where 1 is the index of the westernmost column and index numbers increase by one for each column moving east;

9-12 = Land Use and Land Cover code;

13-16 = Political unit code;

17-20 = Hydrologic unit code;

21-24 = Census county subdivision or SMSA tract code;

25-28 = Federal land ownership code;

29-32 = State land ownership code; and

33-52 = Null (binary zeros) field, if present.

If a given overlay has not been digitized, the codes for that overlay will all be zero. To be sure that a regular grid of cells (forming a UTM rectangle) covers the entire base map quadrangle, a "buffer zone" of cells with all zero attributes has been included in the binary CTG data file.

Binary CTG Map Header

The CTG map header associated with a binary CTG data file is stored in a physically separate sequential file. The header consists of six 32-byte logical records. For the first four records, each 32-byte binary record is equivalent to an 80-character CTG map header card image; each integer in a 5-digit character field is stored in a 2-byte binary integer field, and each integer in a 10-digit character field is stored in a 4-byte binary integer field. The fifth card image header record (with text data) is represented as the fifth and sixth 32-byte binary records with EBCDIC coded characters (the last 16 characters of the card image record are always blank).

Listing of CTG Map Header Data

COMPOSITE THEME GRID CHARACTER DATA OUTPUT:

C T G B T A RUN: JUNE 3, 1982 TIME 19:23:06

GRID CELL MAP HEADER INFORMATION:

TITLE: LAWRENCE, MO KS 1:250,000 QUAD LU PB CN HU FO SO

FILE CREATION DATE: 81084 TIME 0: 0

MAP TYPE: 77   PROJECTION: 1   SCALE 1: 7874016   MAP DATE: 1973

NUMBERS OF FILE ELEMENTS:

CATEGORIES	CELLS	ROWS	COLUMNS	ZONE NUMBER	WEST & NORTH EASTING	EDGES: NORTHING
6	485368	575	884	15	236900	4321100

DUPLICATE POINT TOLERANCE = 0   CELL SIZE IN METERS = 200

MIN COL = 1   MIN ROW = 1   MAX COL = 884   MAX ROW = 575

CONTROL POINT INFORMATION:

	LONGITUDE	LATITUDE	COL	ROW
SOUTH WEST	960000	380000	-1	557
NORTH WEST	960000	390000	17	2
NORTH EAST	940000	390000	883	21
SOUTH EAST	940000	380000	877	576

CHARACTERISTICS OF THE CHARACTER CTG FILE:

THE FILE CONTAINS ONLY GRID CELL (AND NO HEADER) RECORDS.

THE FILE CONSISTS OF 80 CHARACTER RECORDS, ONE GRID CELL PER RECORD.

UTM ZONE, EASTING, AND NORTHING VALUES ARE PART OF EACH CTG DATA RECORD AS THE FIRST THREE INTEGERS, RIGHT JUSTIFIED IN BYTES 1-3, 4-11, AND 12-19.

BYTES 21 80 OF EACH RECORD CONTAIN THE USGS 10 DIGIT INTEGER CODES, RIGHT JUSTIFIED WITHIN 10 BYTE FIELDS, FROM THE FOLLOWING OVERLAYS, IN ORDER:

LAND USE/LAND COVER, POLITICAL UNIT, HYDROLOGIC UNIT, CENSUS SUBDIVISION/TRACT, FEDERAL LAND OWNERSHIP, AND STATE LAND OWNERSHIP.

ONLY RECORDS WITH AT LEAST ONE NON ZERO ATTRIBUTE ARE PART OF THE FILE.
(A VARIABLE NUMBER OF RECORDS EXIST FOR A GIVEN ROW OR COLUMN.)

Sample "Standard" Character-Formatted CTG Data file (without header)

15	240200	4321000	21	0	0	0	0	0
15	240400	4321000	21	0	0	0	0	0
15	240600	4321000	21	0	0	0	0	0
15	240800	4321000	21	0	0	0	0	0
15	241000	4321000	21	0	0	0	0	0
15	241200	4321000	21	0	0	0	0	0
15	240200	4320800	21	201971	0270102	0	2099	2099
15	240400	4320800	21	20197	10270102	0	2099	2099
15	240600	4320800	21	20197	10270102	0	0	2099
15	240800	4320800	21	20197	10270102	0	0	2099
15	241000	4320800	21	20197	10270102	0	0	2099
15	241200	4320800	21	20197	10270102	0	0	2099
15	241400	4320800	21	20197	10270102	0	0	2099
15	241600	4320800	21	0	0	0	0	2099
15	241800	4320800	21	0	0	0	0	2099
15	242000	4320800	21	0	0	0	0	2099
15	242200	4320800	21	0	0	0	0	0
15	242400	4320800	31	0	0	0	0	0
15	242600	4320800	31	0	0	0	0	0
15	242800	4320800	31	0	0	0	0	0
15	243000	4320800	31	0	0	0	0	0
15	243200	4320800	31	0	0	0	0	0
15	1243400	4320800	31	0	0	0	0	0
15	243600	4320800	31	0	0	0	0	0
15	243800	4320800	31	0	0	0	0	0
15	244000	4320800	31	0	0	0	0	0

15	244200	4320800	31	0	0	0	0	0
15	244400	4320800	31	0	0	0	0	0
15	244600	4320800	31	0	0	0	0	0
15	244800	4320800	31	0	0	0	0	0
15	245000	4320800	31	0	0	0	0	0
15	245200	4320800	31	0	0	0	0	0
15	245400	4320800	31	0	0	0	0	0
15	245600	4320800	31	0	0	0	0	0
15	245800	4320800	31	0	0	0	0	0
15	246000	4320800	31	0	0	0	0	0
15	246200	4320800	31	0	0	0	0	0
15	246400	4320800	31	0	0	0	0	0
15	246600	4320800	21	0	0	0	0	0
15	246800	4320800	21	0	0	0	0	0
15	247000	4320800	21	0	0	0	0	0
15	247200	4320800	21	0	0	0	0	0
15	247400	4320800	21	0	0	0	0	0
15	247600	4320800	21	0	0	0	0	0
15	247800	4320800	21	0	0	0	0	0
15	248000	4320800	21	0	0	0	0	0
15	248200	4320800	21	0	0	0	0	0
15	240200	4320600	21	20197	10270102	20197025	2099	2099
15	240400	4320600	21	20197	10270102	20197025	2099	2099
15	240600	4320600	21	20197	10270102	20197025	2099	2099
15	240800	4320600	21	20197	10270102	20197025	2099	2099
15	241000	4320600	21	20197	10270102	20197025	2099	2099

[etc]